Asymptotic Analyses for Complex Evolutionary Systems with Markov and Semi-Markov Switching Using Approximation Schemes

To our teachers

Series Editor
Nikolaos Limnios

Asymptotic Analyses for Complex Evolutionary Systems with Markov and Semi-Markov Switching Using Approximation Schemes

Yaroslav Chabanyuk
Anatolii Nikitin
Uliana Khimka

iSTE

WILEY

First published 2020 in Great Britain and the United States by ISTE Ltd and John Wiley & Sons, Inc.

ISTE Ltd
27-37 St George's Road
London SW19 4EU
UK

www.iste.co.uk

John Wiley & Sons, Inc.
111 River Street
Hoboken, NJ 07030
USA

www.wiley.com

Library of Congress Control Number: 2020936535

British Library Cataloguing-in-Publication Data
A CIP record for this book is available from the British Library
ISBN 978-1-78630-556-5

MIX
Paper from responsible sources
FSC www.fsc.org FSC® C013604

Contents

Acronyms

SDE	Stochastic Differential Equation
IPP	Impulse Perturbation Process
MP	Markov Process
SMP	Semi-Markov Process
SAP	Stochastic Approximation Procedure
SOP	Stochastic Optimization Procedure
MRP	Markov Renewal Process
EMRP	Extended Markov Renewal Process
SSPP	Solving of Singular Perturbed Problem

Introduction

The emergence of the theory of stochastic differential equations played a prominent role in the works of S.N. Bernshtain, M.M. Bogolyubov and M.M. Krylov. A systematic study of stochastic differential equations was first carried out by Y.I. Gichman. The concept of random evolution was introduced by Griego and Hersh (1969) and Bellman (1957).

Applications of such a model were derived from the work of Feng (1999), Fleming and Soner (2006), Feng and Kurtz (2006), which were stimulated by the problems of the stability of stochastic systems.

In the 1960s and 1970s, the problems associated with the theory of random evolution were actively investigated by American mathematicians R. Hersch, M. Pinskii, G. Papanikolau, T. Kurtts, R. Griego, L. Horossey (Skorokhod 1989; Sviridenko 1998; Skorokhod *et al.* 2002; Samoilenko *et al.* 2017) and others. In particular, G. Papanikolaou, D. Stroock and S. Varadan proposed a martingale approach for the proof of boundary theorems (Papanicolaou *et al.* 1977), Stroock using methods similar to solving the singular perturbation problem.

An effective means to prove the limit theorems in the theory of random evolution was developed by Korolyuk and Turbin (1993). Thanks to their methods, A.F. Turbin, O.S. Hochhel and other authors proved important boundary theorems on regular approximations for singularly perturbed solutions of differential equations for different stochastic models, as well as for dynamical systems with coefficients dependent on Markov processes (Feng and Kurtz 2006). Skorokhod (1989), V.E. Shapiro and V.M. Login investigated several important problems of the theory of dynamical systems described by stochastic differential equations. The problem of the stability of dynamical systems with random disturbance of their parameters was investigated by Gorostiza (1972, 1973a, 1973b), E.F. Tsar'kov and V.S. Korolyuk. Problems of stability and of the stochastic approximation of evolution systems with switches were investigated by Y.M. Chabanyuk and in the works of Kurtz (1973,

1987), Donsker and Varadhan (1975a), Papanicolaou (1975), Korolyuk and Turbin (1993) and Koroliuk *et al.* (2010a). V.S. Korolyuk and A.V. Svishchuk also developed the theory of semi-Markovic random evolution, based on the theory of martingales. Many applications of switching processes to the analysis of network systems can be found in the works of V.V. Anisimov (Bainov and Simeonov 1992; Albeverio *et al.* 2008, 2009; Anisimov 2008). Samoilenko I.V. (Albeverio *et al.* 2008, 2009) studied the Markov processes with impulse disturbances in approximation schemes.

Methods of studying the asymptotic properties used in the dissertation have been described in the monographs of Korolyuk and Limnios (2005) and Skorokhod. These methods are mainly applied to models in averaging and diffusion approximation schemes (Chabanyuk *et al.* 2007). At the same time, with a growing time interval, the averaging scheme demonstrates the deterministic averaged behavior of the system, and the scheme of diffusion approximation is stochastic fluctuations around a deterministic averaged trajectory. These two schemes differ in the normalization of the switching process, namely, in the case of the averaging scheme, the acceleration of time is considered by the parameter ε^{-1}, whereas in the case of diffusion approximation, it is considered by the parameter ε^{-2}. An important element of the algorithm is the assumption that the switching process is ergodic.

Note that the main ideas of the random process have been presented in the works of Cramer (1938), Cole (1951), Chernoff (1952), Bellman (1957), Doléans-Dade and Meyer (1970), Papanicolaou (1971a, 1971b, 1973), Hersh (1974), Roth (1974), Cinlar *et al.* (1980) and Doob (1990). Works on the problem of large deviations should be singled out: Freidlin (1978), Ethier and Kurtz (1985), Deuschel and Stroock (1989), Bryc (1990), Dembo and Zeitouni (1993), Mogulskii (1993), O'Brien and Vervaat (1995), Dupuis and Ellis (1997) and Koroliuk and Samoilenko (2014).

Tasks of weak convergence of the random process have been presented in the works of Kushner (1990) and Kolesnik (2001, 2003).

The application of stochastic differential equations has been presented in the following works: Hopf (1950), Pinsky (1968), Papanicolaou and Kellek (1971), Papanicolaou and Hersh (1972), Nashed (1976), Papanicolaou *et al.* (1977), Crandall and Lions (1983), Fukushima and Stroock (1986), Protter (1990), Puhalskii (1991), Davis (1993), Dellacherie and Meyer (1993), Devooght (1997), Hillen (2001), Othmer and Hillen (2002), Hersh (2003), Jacod and Shiryaev (2003) and Orsingher and Sommella (2004).

In the monographs of Korolyuk and Korolyuk (1999) and Limnios and Oprişan (2001), the classification of stochastic models has been considered. In the works of Korolyuk and Chabanyuk (2002) and Chabanyuk (2007), stability conditions of the dynamic system with semi-Markov switchings have been established in the average scheme and the diffusion approximation scheme. The fluctuations of such system have been considered in the paper by Chabanyuk *et al.* (2007).

A natural generalization of the stochastic approximation procedure (Nevelson and Khasminsky 1972) is an examination of the case where the regression function is directly dependent on an external environment with Markov or semi-Markov switching (Chabanyuk 2004a, 2004b, 2019). For the stochastic optimization procedure (Khimka and Chabanyuk 2013), sufficient conditions are obtained for convergence with Markov switching. The properties of the compensating operator of random evolution are used in the case of semi-Markov switching (Sviridenko 1998).

The stochastic approximation procedure has been investigated in the case of impulse perturbing in the work of Khimka and Chabanyuk (2013).

The conditions of asymptotic normality of the stochastic approximation procedure have been established in the works of Chabanyuk (2004a, 2004b, 2006). This property has been considered for the control task in the work of Nikitin and Khimka (2017).

Korolyuk and Limnios (2005) also considered several models for stochastic evolutions by taking into account impulse influences in the Poisson and Levy approximation schemes.

They also presented another important direction of research, namely the study of the dissipative nature of random evolutions under the Levy and Poisson approximation conditions (Bertoin 1996; Koroliuk *et al.* 2010a, 2010b). In our research, an asymptotic dissipative analysis of stochastic differential equations with impulse influences is carried out under the conditions of nonclassical approximation schemes (i.e. Levy and Poisson).

Another important direction of the monograph research is the establishment of conditions for a weak convergence of the diffusion transfer process with Markov switches and the control of the function of the quality criterion with the equilibrium point for which the stochastic approximation procedure is constructed in the series scheme. Under these conditions, the normalized process is also constructed and its asymptotic normality is established in the Ornstein–Uhlenbeck process when the transfer process changes under the influence of the Markov switching on the trajectory of a new evolution from the state in which it was at the moment of switching. The problem of normalized control with a single point of equilibrium of the quality criterion of the diffusion transfer process is studied under the conditions of the stochastic approximation procedure. A similar statement of the problem is completely new and allows for some significant generalizations.

The thesis contains some applications of the developed methods to the analysis of the model of counteraction to information attacks. The monograph includes the following new results:

– under the Levy approximation conditions, a limiting generator was constructed and an asymptotic behavior analysis of stochastic evolution systems with impulse perturbation and Markov switches was carried out;

– under the Poisson approximation conditions, an asymptotic analysis of the behavior of stochastic evolution systems with impulse disturbance and Markov switches was carried out;

– for a Poisson approximation, the boundary process contains two components: deterministic shift and Poisson jump component. In contrast, for a Levy approximation, the Levy limiting process contains three components: deterministic shift, diffusion time and Poisson jump component. In this case, the function that defines a deterministic shift is determined by frequent small jumps of the process; the function, which sets the diffusion component, is determined by frequent large jumps of the process; finally, the high jumps of the boundary process are set by the measure of rare large jumps of the border process;

– an asymptotic dissipative analysis of stochastic differential equations with impulse effects under the conditions of nonclassical approximation schemes (Levy and Poisson) was carried out;

– the conditions of weak convergence of the diffusion transfer process with Markov switches and the control of the function of the quality criterion with the equilibrium point for which the stochastic approximation procedure was constructed in the series scheme;

– for the process of transfer with Markov switches and control under the conditions of existence of a single point of equilibrium of the quality criterion, a normalized process was constructed and its asymptotic normality was established in the Ornstein–Uhlenbeck process when the transfer process changed under the influence of the Markov switching on the trajectory of a new evolution from the state in which it was at the moment of switching;

– for the stochastic approximation procedure, a generalization to the case of the extrinsic environment dependent regression function with Markov and Semi-Markov switches is considered. Sufficient stochastic convergence conditions for Markov switches were obtained for the stochastic optimization procedure.

1

Average Scheme and Diffusion Approximation Scheme

1.1. Stability of stochastic systems in the average scheme

In this section, we analyze the stability of a dynamical system with semi-Markov switchings using a compensating operator of the semi-Markov process. The problem of the stability of a semi-Markov stochastic system is similar to the problem with Markov switchings.

We study the stochastic evolutionary equation

$$du^\varepsilon(t)/dt = C(u^\varepsilon(t), x(t/\varepsilon)), u^\varepsilon(0) = u_0, \quad [1.1]$$

which defines an initial dynamical system with semi-Markov switchings in the series scheme with a small parameter $\varepsilon > 0$, $R^d : u^\varepsilon(t) = (u_k^\varepsilon(t); k = \overline{1,d})$ – Euclidean space. The velocities of the dynamical system are determined by the function vectors $C(u,x) = (C_k(u,x),\ k = \overline{1,d})$, which satisfy the conditions of the existing global solutions for deterministic systems

$$\frac{du_x^\varepsilon(t)}{dt} = C(u_x^\varepsilon(t), x), x \in X, \quad [1.2]$$

where x is the semi-Markov process in the standard phase space $(X, \boldsymbol{X})$ (Korolyuk and Limnios 2005). The average dynamical system is determined by the solution of the evolutionary deterministic equation

$$d\widehat{u}(t)/dt = C(\widehat{u}(t)), \widehat{u}(t) = u_0, \quad [1.3]$$

where the averaging velocity is given by the equality

$$C(u) = \int_X \pi(dx) C(u,x). \quad [1.4]$$

Under the conditions of the stability of the averaged system [1.3], the problem is to establish additional conditions to ensure the stability of the initial system [1.1] with sufficiently small values of the parameter $\varepsilon > 0$.

THEOREM 1.1.– *Assume that for the average dynamical system [1.3] and [1.4], the Lyapunov function $V(u), u \in R^d$ exists, which satisfies the following conditions:*

$C1$: *exponential stability*

$$C(u)V'(u) \leq -c_0 V(u), c_0 > 0;$$

additional conditions

$$C2 : |C(u,x)V'(u)| \leq c_1 V(u);$$

$$C3 : |C(u,x)[C(u,x)V'(u)]'| \leq c_2 V(u);$$

$C4$: *distribution functions $G_x(t), t \geq 0, x \in X$, of the time of stay in the states of the semi-Markov process $x(t), t \geq 0$, satisfy the Cramer condition, uniformly at $x \in X$:*

$$\sup_{x \in X} \int_0^\infty e^{ht} \overline{G}_x(t) dt \leq H < +\infty, h > 0,$$

$C5$: *estimates hold true:*

$$0 < \underline{m} \leq g(x) \leq \overline{m} < +\infty.$$

Then, for all $\varepsilon \leq \varepsilon_0$ (where ε_0 is small enough), the solution of the evolutionary equation [1.1] at all initial conditions $|u^\varepsilon(0)| \leq u^*$ (where u^* is small enough) is asymptotically stable with probability 1:

$$P\{\lim_{t \to \infty} \|u^\varepsilon(t)\| = 0\} = 1. \quad [1.5]$$

Under random initial conditions, the inequality $|Eu^\varepsilon(0)| \leq u^*$ is satisfied, where u^* is small enough.

The extended Markov renewal process (EMRP) has the form

$$u_n^\varepsilon = u^\varepsilon(\tau_n^\varepsilon), x_n^\varepsilon = x(\tau_n^\varepsilon), \tau_n^\varepsilon = \varepsilon \tau_n, n \geq 0. \quad [1.6]$$

DEFINITION 1.1.– *The compensating operator of the EMRP [1.6] is determined by the relation Sviridenko (1998)*

$$\boldsymbol{L}^\varepsilon \varphi(u,x) := \varepsilon^{-1} q(x) E[\varphi(u_{n+1}^\varepsilon, x_{n+1}^\varepsilon,) - \varphi(u,x)|$$

$$u_n^\varepsilon = u, x_n^\varepsilon = x, \tau_n^\varepsilon = t] \quad [1.7]$$

We assume that the family of semigroups $\boldsymbol{C}_t(x)$, $t \geq 0$, $x \in X$, is generated by the accompanying system [1.2], i.e. it is determined by the generator

$$\boldsymbol{C}(x)\varphi(u) = C(u,x)\varphi'(u). \tag{1.8}$$

LEMMA 1.1.– *The compensating operator of the EMRP [1.6] has the analytical form*

$$\boldsymbol{L}^{\varepsilon}\varphi(u,x) = \varepsilon^{-1}q(x)[\int_0^{\infty} G_x(ds)\boldsymbol{C}_{\varepsilon s}(x)\int_X P(x,dy)\varphi(u,y) - \varphi(u,x)].$$

PROOF.– We begin with the calculation of the conditional expectation using semigroups $\boldsymbol{C}_t(x)$. Since semigroups $\boldsymbol{C}_t(x)$ on the Banach space $C(R^d)$ of continuous real-valued functions $\varphi(u), u \in R^d$, with the supremum norm is determined by the relation

$$\boldsymbol{C}_{t'}(x) := \varphi(u(t+t')),$$

under the condition $t' = \theta_x^{\varepsilon} = \varepsilon\theta_x$, we have

$$\boldsymbol{C}_{\theta_x^{\varepsilon}}(x)\varphi(u) = \varphi(u_{n+1}^{\varepsilon}),$$

i.e.

$$\boldsymbol{C}_{\varepsilon\theta_x}(x)\varphi(u) = \varphi(u_{n+1}^{\varepsilon}).$$

From the last relation, we have the equality for the conditional expectation

$$E[\varphi(u_{n+1}^{\varepsilon})|u_n^{\varepsilon} = u, x_n^{\varepsilon} = x, \tau_n^{\varepsilon} = t] = E_{u,x,t}\boldsymbol{C}_{\varepsilon\theta_x}(x)\varphi(u).$$

Using the representation of a semi-Markov kernel

$$Q(x,dy,dt) = G_x(dt)P(x,dy),$$

we obtain

$$E[\varphi(u_{n+1}^{\varepsilon}, x_{n+1}^{\varepsilon})|u_n^{\varepsilon} = u, x_n^{\varepsilon} = x, \tau_n^{\varepsilon} = t]$$

$$E_{u,x,t}\varphi(u_{n+1}^{\varepsilon}, x_{n+1}^{\varepsilon}) = \int_0^{\infty} G_x(ds)\boldsymbol{C}_{\varepsilon s}(x)\int_X P(x,dy)\varphi(u,y).$$

Setting the last representation in [1.7], we have the equality that completes the proof of the lemma.

LEMMA 1.2.– *The compensating operator on test functions* $\varphi(u,x) \in C^2(R^d \times X)$ *has the asymptotic representation*

$$\boldsymbol{L}^{\varepsilon}\varphi(u,x) = \varepsilon^{-1}Q\varphi(u,x) + \theta_1^{\varepsilon}(x)\varphi(u,x), \tag{1.9}$$

and

$$\boldsymbol{L}^{\varepsilon}\varphi(u,x) = \varepsilon^{-1}Q\varphi(u,x) + \boldsymbol{C}(x)\boldsymbol{P}\varphi(u,x) + \varepsilon\theta(x)\varphi(u,x), \qquad [1.10]$$

where

$$Q\varphi(\cdot,x) = q(x)[\boldsymbol{P}-\boldsymbol{I}]\varphi(\cdot,x), x \in X, \boldsymbol{P}\varphi(\cdot,x) = \int_X P(x,dy)\varphi(\cdot,y).$$

The reminder operators $\theta^{\varepsilon}(x)$ θ_1^{ε} have the representation

$$\theta(x)\varphi(u,x) = q(x)\boldsymbol{C}^2(x)\boldsymbol{C}_2^{\varepsilon}(x)\boldsymbol{P}\varphi(u,\cdot),$$
$$\theta_1^{\varepsilon}(x)\varphi(u,x) = q(x)\boldsymbol{C}(x)\boldsymbol{C}_1^{\varepsilon}(x)\boldsymbol{P}\varphi(u,\cdot), \qquad [1.11]$$

where

$$\boldsymbol{C}_2^{\varepsilon}(x)\varphi(u) := \int_0^{\infty} \overline{G}_x^{(2)}(s)\boldsymbol{C}_{\varepsilon s}(x)\varphi(u)ds,$$

$$\boldsymbol{C}_1^{\varepsilon}(x) = \int_0^{\infty} \overline{G}_x(s)\boldsymbol{C}_{\varepsilon s}(x)ds,$$

$$\overline{G}_x^{(2)}(s) := \int_s^{\infty} \overline{G}_x(t)dt.$$

PROOF.– The proof is based on the following semigroup representation:

$$\boldsymbol{C}_{\varepsilon s}(x) = \boldsymbol{I} + \varepsilon s\boldsymbol{C}(x) + \varepsilon^2\boldsymbol{C}^2(x)\boldsymbol{T}_s^{\varepsilon}(x) = \boldsymbol{I} + \boldsymbol{C}(x)\int_0^{\varepsilon^2 s} \boldsymbol{C}_v(x)dv,$$

where

$$\boldsymbol{T}_s^{\varepsilon}(x) = \int_0^s (s-v)\boldsymbol{C}_{\varepsilon v}(x)dv.$$

We note that for calculation of the remainder operators, we use the following relations:

$$\boldsymbol{C}_2^{\varepsilon}(x) = \int_0^{\infty} G_x(ds)\boldsymbol{C}_s^{\varepsilon}(x) = \int_0^{\infty} \overline{G}_x^{(2)}(s)\boldsymbol{C}_{\varepsilon s}(x)ds,$$

$$\boldsymbol{C}_1^{\varepsilon}(x) = \int_0^{\infty} \overline{G}_x(s)\boldsymbol{C}_{\varepsilon s}(x)ds.$$

We also note that

$$\int_0^\infty \overline{G}_x^{(2)}(s)ds = g_2(x)/2, g_2(x) := E\theta_x^2 = \int_0^\infty t^2 G_x(dt).$$

Let us consider the perturbed Lyapunov function:

$$V^\varepsilon(u,x) = V(u) + \varepsilon V_1(u,x), \qquad [1.12]$$

where $V(u), u \in R^d$, is the Lyapunov function for the average system [1.3], under conditions $C1 - C3$ of the theorem.

The perturbation $V_1(u,x), u \in R^d, x \in X$, is determined by the solution of the singular perturbation problem for the operator

$$\boldsymbol{L}_0^\varepsilon \varphi(u,x) = \varepsilon^{-1} Q\varphi(u,x) + \boldsymbol{C}(x)\varphi(u,x). \qquad [1.13]$$

The solution of the singular perturbation problem for the operator [1.13] is determined by the relation

$$\boldsymbol{L}_0^\varepsilon V^\varepsilon(u,x) = \boldsymbol{C}V(u) + \varepsilon \boldsymbol{C}(x)V_1(u,x),$$

$$\boldsymbol{C}V(u) = C(u)V'(u),$$

$$V_1(u,x) = R_0\widetilde{\boldsymbol{C}}(x)V(u),$$

where $\widetilde{\boldsymbol{C}}(x) := \boldsymbol{C}(x) - \boldsymbol{C}$ and the operator $\boldsymbol{R}_0 = \Pi - (Q+\Pi)^{-1}$ is a potential of the operator Q. We know that under condition $C4$ of the theorem, the operator $\boldsymbol{R}_0$ is bounded.

LEMMA 1.3.– *The solution of the singular perturbation problem for the compensating operator [1.10] on the perturbed Lyapunov function [1.12] is determined by the relations:*

$$\boldsymbol{L}^\varepsilon V^\varepsilon(u,x) = \boldsymbol{C}V(u) + \varepsilon\theta_0^\varepsilon(x)V(u). \qquad [1.14]$$

The reminder operator has the form:

$$\theta_0^\varepsilon(x)V(u) = q(x)[\boldsymbol{C}^2(x)\boldsymbol{C}_2^\varepsilon(x) + \boldsymbol{C}(x)\boldsymbol{C}_1^\varepsilon(x)\boldsymbol{P}R_0\widetilde{\boldsymbol{C}}(x)]V(u). \qquad [1.15]$$

PROOF.– We use two representations of the generator $\boldsymbol{L}^\varepsilon$ of Lemma 1.2, respectively for [1.10]

$$\boldsymbol{L}_{(0)}^\varepsilon := \varepsilon^{-1}Q + \theta_1^\varepsilon(x),$$

and for [1.11]

$$\boldsymbol{L}_{(1)}^\varepsilon := \varepsilon^{-1}Q + \boldsymbol{C}(x)\boldsymbol{P} + \varepsilon\theta^\varepsilon(x).$$

We represent the generator $\boldsymbol{L}^{\varepsilon}$ on the perturbed Lyapunov function [1.12] as

$$\boldsymbol{L}^{\varepsilon}V^{\varepsilon}(u,x) = \varepsilon\boldsymbol{L}^{\varepsilon}_{(0)}V_1(u,x) + \boldsymbol{L}^{\varepsilon}_{(1)}V(u).$$

Since

$$\varepsilon\boldsymbol{L}^{\varepsilon}_{(0)}V_1(u,x) = QV_1(u,x) + \varepsilon\theta^{\varepsilon}_1(x)V_1(u,x),$$

and

$$\boldsymbol{L}^{\varepsilon}_{(1)}V(u) = \varepsilon^{-1}QV(u) + \boldsymbol{C}(x)V(u) + \varepsilon\theta^{\varepsilon}(x)V(u),$$

then

$$\boldsymbol{L}^{\varepsilon}V^{\varepsilon}(u,x) = \varepsilon^{-1}QV(u) + QV_1(u,x) + \boldsymbol{C}(x)V(u) + \varepsilon[\theta^{\varepsilon}(x)V(u) + \theta^{\varepsilon}_1(x)V_1(u,x)].$$

From the representation of the remainder operators $\theta^{\varepsilon}(x)$ and θ^{ε}_1, as well as the perturbation $V_1(u,x)$ in the form [1.13], we obtain the representation of the remainder term in the form [1.15].

We note that under conditions $C1 - C5$ of the theorem, we have the inequality

$$\boldsymbol{L}^{\varepsilon}V^{\varepsilon}(u,x) \leq -cV(u), \quad [1.16]$$

To prove inequality [1.16], we first estimate the remainder operator [1.15], using conditions $C2 - C3$ of the theorem

$$|\theta^{\varepsilon}_0(x)V(u)| \leq c_0V(u), \quad [1.17]$$

and then estimate the perturbation $V_1(u,x)$:

$$|V_1(u,x)| \leq c_1V(u). \quad [1.18]$$

According to the exponential stability condition $C1$, the first term in [1.14] satisfies the inequality

$$\boldsymbol{C}V(u) = C(u)V'(u) \leq -c_0V(u), c_0 > 0. \quad [1.19]$$

Combining the estimates [1.17], [1.18] and [1.19], we obtain [1.16] for each $\varepsilon \leq \varepsilon_0$, where ε_0 is small enough, at some value of $c > 0$.

The condition $C2$ of the theorem provides an estimate of the semigroups $\boldsymbol{C}_t(x)$, which is generated by the operator [1.8]:

$$|\boldsymbol{C}_t(x)V(u)| \leq e^{bt}V(u). \quad [1.20]$$

The following inequalities also hold true (see [1.12], [1.17] and [1.18]):

$$0 \leq b_1V^{\varepsilon}(u,x) \leq V(u) \leq b_2V^{\varepsilon}(u,x), 0 < b_1 < b_2.$$

LEMMA 1.4.– *Extended RP [1.6] is characterized by a martingale*

$$\mu_{n+1}^{\varepsilon} = \varphi(u_{n+1}^{\varepsilon}, x_{n+1}^{\varepsilon}, \tau_{n+1}^{\varepsilon}) - \sum_{k=0}^{n} \theta_{k+1} \boldsymbol{L}^{\varepsilon} \varphi(u_k^{\varepsilon}, x_k^{\varepsilon}, \tau_k^{\varepsilon}), n \geq 0. \qquad [1.21]$$

PROOF.– The martingale property of a consequence $(\mu_{n+1}^{\varepsilon}, n \geq 0)$ yields from the uniformity of the compensating operator:

$$\begin{aligned} &E[\varphi(u_{n+1}^{\varepsilon}, x_{n+1}^{\varepsilon}, \tau_{n+1}^{\varepsilon}) - \varphi(u_n^{\varepsilon}, x_n^{\varepsilon}, \tau_n^{\varepsilon}) | F_n^{\varepsilon}] \\ &= E[\varphi(u_1^{\varepsilon}, x_1^{\varepsilon}, \tau_1^{\varepsilon}) - \varphi(u_0^{\varepsilon}, x_0^{\varepsilon}, \tau_0^{\varepsilon}) | F_0^{\varepsilon}]. \end{aligned}$$

Here, σ-algebras $F_n^{\varepsilon} := \sigma\{u_k^{\varepsilon}, x_k^{\varepsilon}, \tau_k^{\varepsilon}; 0 \leq k \leq n\}, n \geq 0$, are generated by the EMRP.

We denote

$$\begin{aligned} &\Phi^{\varepsilon}(t) := \varphi(u^{\varepsilon}(\tau^{\varepsilon}(t)), x^{\varepsilon}(\tau^{\varepsilon}(t)), \tau^{\varepsilon}(t)), \\ &\Phi_{+}^{\varepsilon}(t) := \varphi(u^{\varepsilon}(\tau_{+}^{\varepsilon}(t)), x^{\varepsilon}(\tau_{+}^{\varepsilon}(t)), \tau_{+}^{\varepsilon}(t)), \\ &\tau^{\varepsilon}(t) := \tau_{\nu^{\varepsilon}(t)}^{\varepsilon}, \tau_{+}^{\varepsilon}(t) := \tau_{\nu_{+}^{\varepsilon}(t)}^{\varepsilon}, \nu_{+}^{\varepsilon}(t) := \nu^{\varepsilon}(t) + 1, \nu^{\varepsilon}(t) := \nu(\varepsilon t). \end{aligned}$$

Note that $\nu_{+}^{\varepsilon}(t)$ is the Markovian moment for the flow F_n^{ε}.

Furthermore, we use the martingale property of the process with continuous time:

$$\zeta^{\varepsilon}(t) = \Phi_{+}^{\varepsilon}(t) - \int_{0}^{\tau_{+}^{\varepsilon}(t)} \boldsymbol{L}^{\varepsilon} \Phi^{\varepsilon}(s) ds. \qquad [1.22]$$

We note that

$$\zeta^{\varepsilon}(\tau_n^{\varepsilon}) = \mu_{n+1}^{\varepsilon}, n \geq 0,$$

and

$$\zeta^{\varepsilon}(t) = \zeta^{\varepsilon}(\tau_{+}^{\varepsilon}(t)), \tau^{\varepsilon}(t) < t < \tau_{+}^{\varepsilon}(t).$$

We first prove the following lemma, which is based on proving the stability of the dynamical system.

LEMMA 1.5.– *The process*

$$\zeta_c^{\varepsilon}(t) = e^{c\tau_{+}^{\varepsilon}(t)} \Phi_{+}^{\varepsilon}(t) - \int_{0}^{\tau_{+}^{\varepsilon}(t)} [c e^{cs} \Phi_{+}^{\varepsilon}(s) + e^{c\tau^{\varepsilon}(s)} \boldsymbol{L}^{\varepsilon} \Phi^{\varepsilon}(s)] ds \qquad [1.23]$$

has the martingale property:

$$E[\zeta_c^\varepsilon(t) - \zeta_c^\varepsilon(s)|F_s^\varepsilon] = 0, 0 \leq s < t,$$

with respect to the flow of σ-algebras

$$F_s^\varepsilon := \{\sigma(u^\varepsilon(l)), x^\varepsilon(l), \tau^\varepsilon(l); 0 \leq l \leq s\},$$

for any fixed real value of the parameter $c \in R$.

PROOF.– The process [1.23] can be represented in an equivalent form:

$$\zeta_c^\varepsilon(t) = e^{c\tau_+^\varepsilon(t)}\zeta^\varepsilon(t) - \int_0^{\tau_+^\varepsilon(t)} ce^{cs}\zeta^\varepsilon(s)ds, \qquad [1.24]$$

where $\zeta^\varepsilon(t), t \geq 0$, is a process that is given by formula [1.22].

We also note that the martingale property of the process [1.24] follows from the fact that the process is jumpy and takes values:

$$\zeta_c^\varepsilon(\tau_n^\varepsilon) = e^{c\tau_{n+1}^\varepsilon}\mu_{n+1}^\varepsilon - \sum_{k=0}^{n}[e^{c\tau_{k+1}^\varepsilon} - e^{c\tau_k^\varepsilon}]\mu_{k+1}^\varepsilon, \qquad [1.25]$$

i.e.

$$\zeta_c^\varepsilon(\tau_n^\varepsilon) = \mu_0 + \sum_{k=0}^{n} e^{c\tau_k^\varepsilon}[\mu_{k+1}^\varepsilon - \mu_k^\varepsilon], n \geq 0. \qquad [1.26]$$

To prove the equivalence of representations [1.23] and [1.24], we calculate a double integral:

$$\begin{aligned}
I^\varepsilon(t) &= \int_0^{\tau_+^\varepsilon(t)} ce^{cs}ds \int_0^{\tau_+^\varepsilon(s)} \boldsymbol{L}^\varepsilon\Phi^\varepsilon(v)dv \\
&= \sum_{k=0}^{\nu^\varepsilon(t)} \int_{\tau_k^\varepsilon}^{\tau_{k+1}^\varepsilon} ce^{cs}ds \int_0^{\tau_{k+1}^\varepsilon} \boldsymbol{L}^\varepsilon\Phi^\varepsilon(v)dv \\
&= \sum_{k=0}^{\nu^\varepsilon(t)} (e^{c\tau_{k+1}^\varepsilon} - e^{c\tau_k^\varepsilon}) \sum_{r=0}^{k} \varepsilon\theta_{r+1}\boldsymbol{L}^\varepsilon\Phi^\varepsilon(\tau_r^\varepsilon) \\
&= \sum_{r=0}^{\nu^\varepsilon(t)} \theta_{r+1}\boldsymbol{L}^\varepsilon\Phi^\varepsilon(\tau_r^\varepsilon)(e^{c\tau_{r+1}^\varepsilon} - e^{c\tau_r^\varepsilon}).
\end{aligned}$$

Finally, we obtain

$$I^{\varepsilon}(t) = e^{c\tau_{+}^{\varepsilon}(t)} \int_{0}^{\tau_{+}^{\varepsilon}(t)} \boldsymbol{L}^{\varepsilon}\Phi^{\varepsilon}(s)ds - \int_{0}^{\tau_{k+1}^{\varepsilon}} e^{c\tau^{\varepsilon}(s)}\boldsymbol{L}^{\varepsilon}\Phi^{\varepsilon}(s)ds.$$

Thus, we obtain representation [1.23] from [1.22] and [1.24].

Furthermore, we analyze the process

$$\eta^{\varepsilon}(t) = \int_{0}^{\tau_{+}^{\varepsilon}(t)} [e^{cs}cV_{+}^{\varepsilon}(s) + e^{c\tau^{\varepsilon}(t)}\boldsymbol{L}^{\varepsilon}V^{\varepsilon}(s)]ds. \qquad [1.27]$$

LEMMA 1.6.– *Under the theorem conditions, the process* $\eta^{\varepsilon}(t), t \geq 0$, *satisfies the relation*

$$E[\eta^{\varepsilon}(t) - \eta^{\varepsilon}(s)|F_{s}^{\varepsilon}] \leq 0, 0 \leq s < t. \qquad [1.28]$$

for all the values of the parameter c and for each $\varepsilon \leq \varepsilon_0$ *(which is small enough).*

PROOF.– Since the process [1.27] is jumping, it is sufficient to consider the conditional expectation at moments of Markov renewals:

$$\eta_{n+1}^{\varepsilon} := \eta^{\varepsilon}(\tau_{n}^{\varepsilon}) = \int_{0}^{\tau_{n+1}^{\varepsilon}} [e^{cs}cV_{+}^{\varepsilon}(s) + e^{c\tau^{\varepsilon}(s)}\boldsymbol{L}^{\varepsilon}V^{\varepsilon}(s)]ds, \qquad [1.29]$$

i.e.

$$\eta_{n+1}^{\varepsilon} = \sum_{k=0}^{n} e^{c\tau_{k}^{\varepsilon}}\alpha_{k+1}^{\varepsilon}, \qquad [1.30]$$

where

$$\alpha_{k+1}^{\varepsilon} = \beta_{k+1}^{\varepsilon} + \varepsilon\gamma_{k+1}^{\varepsilon}, \qquad [1.31]$$

$$\beta_{k+1}^{\varepsilon} = (e^{\varepsilon c\theta_{k+1}} - 1)V_{k+1}^{\varepsilon},$$

$$\gamma_{k+1}^{\varepsilon} = \varepsilon\theta_{k+1}\boldsymbol{L}^{\varepsilon}V_{k}^{\varepsilon},$$

$$V_{n}^{\varepsilon} := V^{\varepsilon}(u_{n}^{\varepsilon}, x_{n}^{\varepsilon}, \tau_{n}^{\varepsilon}).$$

For estimating the conditional expectation of the first summand in [1.31], we use the estimation of the semigroup $\boldsymbol{C}_t(x)$ (see [1.20]), which is generated by the operator [1.8],

$$|\boldsymbol{C}_t(x)V(u)| \leq e^{bt}V(u), b > 0.$$

We can now obtain

$$
\begin{aligned}
E[\beta_{k+1}^\varepsilon | F_k^\varepsilon] &= E[(e^{\varepsilon c \theta_{k+1}} - 1)V_{k+1}^\varepsilon | F_k^\varepsilon] \\
&= E[(e^{\varepsilon c \theta_{k+1}} - 1)\boldsymbol{C}_\varepsilon \theta_{k+1}(x_k) V_k^\varepsilon | F_k^\varepsilon] \\
&\leq E[(e^{\varepsilon c \theta_{k+1}} - 1)e^{\varepsilon c \theta_{k+1}} | F_k^\varepsilon] V_k^\varepsilon \\
&= E[(e^{\varepsilon (c+b)\theta_{k+1}} - e^{\varepsilon b \theta_{k+1}}) | F_k^\varepsilon] V_k^\varepsilon \\
&= \varepsilon[b(g_k^\varepsilon(b+c) - g_k^\varepsilon(b)) + c g_k^\varepsilon(b+c)] V_k^\varepsilon .
\end{aligned}
$$

From the equality

$$
g_k^\varepsilon(c) = \int_0^\infty \overline{G}_k(t) e^{\varepsilon c t} dt,
$$

and condition $C4$ of the theorem, we obtain the estimation

$$
m_k \leq g_k^\varepsilon(c) \leq m_k(1 + \delta_\varepsilon)
$$

where $\delta_\varepsilon \leq \delta_0$, δ_0 is small enough when $\varepsilon \leq \varepsilon_0$, where ε_0 is small enough.

Finally, we have the following estimation:

$$
E[\beta_{k+1}^\varepsilon | F_k^\varepsilon] \leq \varepsilon[b\delta_\varepsilon + c m_k(1 + \delta_\varepsilon)] V_k^\varepsilon < \varepsilon \delta_0 V_k^\varepsilon ,
$$

when $\varepsilon \leq \varepsilon_0$ and δ_0 are small enough.

The conditional expectation of the second summand in [1.31] estimates according to inequality [1.19] and the condition

$$
E[\gamma_{k+1}^\varepsilon | F_k^\varepsilon] = m_k \boldsymbol{L}^\varepsilon V_k^\varepsilon \leq -c_0 m_k V_k^\varepsilon \leq -c_0 \underline{m} V_k^\varepsilon .
$$

Finally, we obtain

$$
E[\alpha_{k+1}^\varepsilon | F_k^\varepsilon] \leq -\varepsilon[c_0 \underline{m} - \delta_0] V_k^\varepsilon \leq -\varepsilon c_0 V_k^\varepsilon , c_0 > 0,
$$

when $\delta_0 < c_0 \underline{m}$.

PROOF OF THEOREM 1.1.– First, it follows from Lemma 1.6 that the consequence

$$
w_{n+1}^\varepsilon = e^{c\tau_{n+1}^\varepsilon} V_{n+1}^\varepsilon , V_{n+1}^\varepsilon := V^\varepsilon(u_{n+1}^\varepsilon, x_{n+1}^\varepsilon, \tau_{n+1}^\varepsilon),
$$

is a supermartingale:

$$
E[w_{n+1}^\varepsilon | F_n^\varepsilon] \leq w_n^\varepsilon , n \geq 0. \qquad [1.32]
$$

Indeed, according to Lemma 1.5, we have (see [1.23], [1.29] and [1.30])

$$w_{n+1}^{\varepsilon} = \zeta_c^{\varepsilon}(\tau_n^{\varepsilon}) + \zeta_{n+1}^{\varepsilon}.$$

The martingale property of the consequence [1.26] and the condition [1.28] provides the condition [1.32]:

$$\begin{aligned} E[w_{n+1}^{\varepsilon}|F_n^{\varepsilon}] &= E[\zeta_c^{\varepsilon}(\tau_n^{\varepsilon}) + \zeta_{n+1}^{\varepsilon}|F_n^{\varepsilon}] \\ &= w_n^{\varepsilon} + E[\zeta_c^{\varepsilon}(\tau_n^{\varepsilon}) - \zeta_c^{\varepsilon}(\tau_{n+1}^{\varepsilon})|F_n^{\varepsilon}] + E[\zeta_{n+1}^{\varepsilon} - \zeta_n^{\varepsilon}|F_n^{\varepsilon}] \leq w_n^{\varepsilon}. \end{aligned}$$

The regularity of the Markov renewal process (MRP) means that

$$\tau_n^{\varepsilon} \Rightarrow \infty, n \to \infty,$$

with probability 1. Thus, for the event $A_{T,n}^{\varepsilon} = \{\tau_n^{\varepsilon} > T\}$, we obtain the estimation of the probability

$$P\{A_{T,n}^{\varepsilon}\} \geq 1 - \Delta, n \geq N_T,$$

for each $\Delta > 0$.

Let us denote

$$a_n^{\varepsilon} := e^{c(\tau_n^{\varepsilon} - T)}, n \geq 0.$$

For the event $A_{T,n}^{\varepsilon}$, we have $a_n^{\varepsilon} \geq 1, n \geq 0$.

We can now estimate the probability

$$\begin{aligned} &P\{e^{cT} \sup_{n \geq N_T} V(u_n^{\varepsilon}) > \delta \bigcap A_{T,n}^{\varepsilon}\} \leq P\{e^{cT} \sup_{n \geq N_T} a_n^{\varepsilon} V(u_n^{\varepsilon}) > \delta \bigcap A_{T,n}^{\varepsilon}\} \\ &\leq P\{\sup_{n \geq N_T} e^{c\tau_n^{\varepsilon}} V(u_n^{\varepsilon}) > \delta \bigcap A_{T,n}^{\varepsilon}\}. \end{aligned}$$

Using [1.4], we have

$$P\{e^{cT} \sup_{n \geq N_T} V(u_n^{\varepsilon}) > \delta \bigcap A_{T,n}^{\varepsilon}\} \leq P\{\sup_{n \geq N_T} w_n^{\varepsilon} > \delta \bigcap A_{T,n}^{\varepsilon}\}.$$

The supermartingale property [1.32] gives the following estimation for the probability

$$P\{e^{cT} \sup_{n \geq N_T} V(u_n^{\varepsilon}) > \delta \bigcap A_{T,n}^{\varepsilon}\} \leq V(u)/\delta_2,$$

where $\delta_2 = \delta b_2 / b_1$.

We then obtain

$$P\{e^{cT}\sup_{n\geq N_T}V(u_n^\varepsilon)>\delta\}=P\{e^{cT}\sup_{n\geq N_T}V(u_n^\varepsilon)>\delta\bigcap A_{T,n}^\varepsilon\}$$

$$+P\{e^{cT}\sup_{n\geq N_T}V(u_n^\varepsilon)>\delta\bigcap\overline{A}_{T,n}^\varepsilon\}\leq V(u)/\delta_2+\Delta\leq 2\Delta \qquad [1.33]$$

at $|u|\leq u_0$, which is small enough.

Considering the relations

$$\{e^{cT}\sup_{n\geq N_T}V(u_n^\varepsilon)>\delta\}\subset\{\lim_{n\to\infty}V(u_n^\varepsilon)=0\}=\{\lim_{n\to\infty}\|u_n^\varepsilon\|=0\}$$

we complete the proof of the convergence

$$P\{\lim_{n\to\infty}\|u_n^\varepsilon\|=0\}=1. \qquad [1.34]$$

Theorem 1.1 is the consequence of [1.32] and estimates [1.19] for the semigroup that generates the stochastic system $u^\varepsilon(t)$. Indeed, the estimation follows from condition $C5$ of the theorem:

$$P\{\max_{1\leq n\leq\nu^\varepsilon(T/\varepsilon)}\theta_n>\delta\}\leq C\Delta(\varepsilon), \qquad [1.35]$$

where $\Delta(\varepsilon)\to 0\ \varepsilon\to 0$.

We consider the representation

$$V_T^\varepsilon:=\sup_{0\leq t\leq T}V(u^\varepsilon(t))$$

$$=\sup_{1\leq n\leq\nu^\varepsilon(T/\varepsilon)}[\sup_{0\leq s\leq\theta_n}|\boldsymbol{C}_{\varepsilon s}(x_{n-1}^\varepsilon)V_{n-1}^\varepsilon|].$$

Using the estimation [1.20] for the semigroup, we have the inequality

$$V_T^\varepsilon\leq\sup_{1\leq n\leq\nu^\varepsilon(T/\varepsilon)}e^{\varepsilon c_1\theta_n}V_{n-1}^\varepsilon=:\overline{V}_T^\varepsilon \qquad [1.36]$$

Thus, the estimation

$$P\{\sup_{0\leq t\leq T}V(u^\varepsilon(t))>\delta\}\leq P\{\overline{V}_T^\varepsilon>\delta\}. \qquad [1.37]$$

holds true.

Now for estimating the right-hand side of [1.37], we use the relation

$$P\{\overline{V}_T^\varepsilon>\delta\}=P\{\overline{V}_T^\varepsilon>\delta\bigcap_{1\leq n\leq\nu^\varepsilon(T/\varepsilon)}\max\theta_n>T/\varepsilon\}$$

$$+P\{\overline{V}_T^\varepsilon>\delta\bigcap_{1\leq n\leq\nu^\varepsilon(T/\varepsilon)}\max\theta_n\leq T/\varepsilon\}. \qquad [1.38]$$

From [1.36] and [1.33], we estimate the first summand in [1.38]:

$$P\{\overline{V}_T^\varepsilon > \delta \bigcap \max_{1\le n\le \nu^\varepsilon(T/\varepsilon)} \theta_n \le T/\varepsilon\}$$

$$\le P\{e^{c_1 T} \sup_{1\le n\le \nu^\varepsilon(T/\varepsilon)} V_{n-1}^\varepsilon > \delta\} \le 2\Delta.$$

We estimate the second summand using [1.35]:

$$P\{\overline{V}_T^\varepsilon > \delta \bigcap \max_{1\le n\le \nu^\varepsilon(T/\varepsilon)} \theta_n > T/\varepsilon\} \le C\Delta(\varepsilon),$$

From the last two estimations using [1.37] and [1.38], we obtain

$$P\{\sup_{1\le t\le T} V(u^\varepsilon(t)) > \delta\} \le 2\Delta + C\Delta(\varepsilon)$$

when δ is small enough.

Using the Korolyuk model theorem Korolyuk and Limnios (2005), we obtain the result of Theorem 1.1.

1.2. Stability of stochastic systems in the diffusion approximation scheme

The analysis of the stability of the dynamical system using the diffusion approximation scheme with semi-Markov switchings can be considered in a more general form and determined by the compensating operator of the semi-Markov process.

The asymptotic representation of the compensating operator that is constructed in this section, in fact, allows the problem of stability to be reformulated for a system with semi-Markov switchings to the same problem for a system with Markov switchings.

The dynamical system in the semi-Markov environment at diffusion approximation conditions can be represented by the stochastic evolutionary equation

$$\frac{du^\varepsilon(t)}{dt} = \varepsilon^{-1} C_1(u^\varepsilon(t), x(\frac{t}{\varepsilon^2})) + C_0(u^\varepsilon(t), x(\frac{t}{\varepsilon^2})), \qquad [1.39]$$

$$u^\varepsilon(0) = u_0,$$

where $\varepsilon > 0$ is a small parameter and $u^\varepsilon(t) = (u_k^\varepsilon(t), k = \overline{1, d})$.

Velocities $C_k(u,x) = (C_{ki}(u,x); i = \overline{1,d}\,), k = 1,0, u \in R^d, x \in X$, satisfy all the conditions that provide existing global solutions for deterministic systems for each $\varepsilon > 0$:

$$\frac{du_x^\varepsilon(t)}{dt} = \varepsilon^{-1}C_1(u_x^\varepsilon(t),x) + C_0(u_x^\varepsilon(t),x), x \in X. \qquad [1.40]$$

Here, $x(t), t > 0$, is the semi-Markov process (SMP) in a standard phase state space $(X, \boldsymbol{X})$, which is generated by the Markov renewal process $x_n, \tau_n, n \geq 0$.

We study the stability of the stochastic system [1.39] under the condition of stability for an averaged system:

$$\begin{cases} du(t) = C_0(u(t))dt + d\zeta(t) \\ d\zeta(t) = a(u(t))dt + \sigma(u(t))dw(t). \end{cases} \qquad [1.41]$$

and assume that the balance condition

$$\int_X \pi(dx)C_1(u,x) \equiv 0, u \in R^d,$$

is fulfilled.

The averaging is determined by the stationary distribution $\pi(dx)$:

$$C_0(u) := \int_X \pi(dx)C_0(u,x).$$

The diffusion process $\zeta(t), t \geq 0$, is determined by the drift vector function

$$a(u) = a_1(u) + a_2(u),$$

where

$$a_1(u) = \int_X \pi(dx)C_1(u,x)R_0C_1'(u,x),$$

$$a_2(u) = \frac{1}{2}q\int_X \rho(dx)\mu(x)C_1(u,x)C_1'(u,x), \qquad [1.42]$$

and the variance matrix $\sigma(u)$ is given by the relations:

$$B(u) = \sigma(u)\sigma^*(u),$$

under the positive definite condition of the matrix

$$B(u) = B_0(u) + B_1(u), \qquad [1.43]$$

where

$$B_0(u) = 2\int_X \pi(dx)C_1(u,x)R_0C_1(u,x),$$

$$B_1(u) = q\int_X \rho(dx)\mu(x)C_1^2(u,x). \tag{1.44}$$

From [1.42] and [1.43], we have

$$\mu(x) = g_2(x) - 2g^2(x), g_2(x) = \int_0^\infty \overline{G}_x^{(2)}(s)ds,$$

$\bar{G}_x^{(2)}(t) := \int_t^\infty \bar{G}_x(s)ds.$

For exponential distribution functions with the intensity $q(x) = g^{-1}(x)$, we have $\mu(x) = 0$. Therefore, the terms of the drift $a_2(u)$ and the variance $B_1(u)$ describe the non-Markov switching process.

The generator of the stochastic system [1.41] is determined by the test functions $\varphi(u) \in C^2(R^d)$ from the relation

$$\mathbf{L}\varphi(u) = C(u)\varphi'(u) + \frac{1}{2}Sp[B(u)\varphi''(u)], \tag{1.45}$$

where

$$C(u) = C_0(u) + a(u). \tag{1.46}$$

Under the conditions of convergence of the stochastic system [1.39] to the averaged system [1.41] at $\varepsilon \to 0$, we set additional conditions that provide the stability of the initial system [1.39] for all $\varepsilon \leq \varepsilon_0$, when ε_0 is small enough.

We study the stability of the system [1.39] under exponential stability conditions for the averaged system [1.41]:

$$\frac{d\widetilde{u}(t)}{dt} = C_0(\widetilde{u}(t)).$$

THEOREM 1.2.– *We assume that for the averaged stochastic system [1.41], determined by the generator [1.45], there exists the Lyapunov function $V(u)$, $u \in R^d$ under exponential stability conditions:*

$$C1 : C_0(u)V'(u) \leq -c_0V(u),\ c_0 > 0,$$

and the following additional conditions at $k, r, l = 0, 1$ hold true:

$C2 :| C_k(u,x)V'(u) |\leq c_1 V(u), c_1 > 0,$

$| C_k(u,x)R_0[C_r(u,x)V'(u)]' |\leq c_2 V(u), c_2 > 0,$

$| C_k(u,x)R_0[C_r(u,x)R_0[C_l(u,x)V'(u)]']' |\leq c_3 V(u), c_3 > 0,$

$C3$: *the distribution functions* $G_x(t), t \geq 0, x \in X$, *satisfy the Cramer condition uniformly at* $x \in X : \sup_{x\in X} \int_0^\infty e^{ht}\overline{G}_x(t)dt \leq H < +\infty, h > 0$, *and*

$0 < \underline{m} \leq g(x) \leq \overline{m} \leq +\infty.$

Then, for all $\varepsilon \leq \varepsilon_0$, *when* ε_0 *is small enough, the solution of the evolutionary equation [1.39], at all initial conditions* $| u^\varepsilon(0) |\leq u^*$, u^*, *when* ε *is small enough, is asymptotically stable with probability 1:*

$$P\{\lim_{t\to\infty} \| u^\varepsilon(t) \|= 0\} = 1.$$

The EMRP is determined by the sequence

$$u_n^\varepsilon = u^\varepsilon(\tau_n^\varepsilon), x_n^\varepsilon = x(\tau_n^\varepsilon), \tau_n^\varepsilon = \varepsilon^2 \tau_n, n \geq 0. \quad [1.47]$$

The compensating operator of the EMRP [1.47] is determined by the relation:

$$\mathbf{L}^\varepsilon\varphi(u,x,t) = \varepsilon^{-2} q(x)[E\{\varphi(u_{n+1}^\varepsilon, x_{n+1}^\varepsilon, \tau_{n+1}^\varepsilon) u_n^\varepsilon = u, x_n^\varepsilon = x, \tau_n^\varepsilon = t\} - \varphi(u,x,t)]. \quad [1.48]$$

We consider a set of semigroups $\boldsymbol{C}_t^\varepsilon(x)$, $t \geq 0, x \in X$, that is generated by the accompanying system [1.40], which can be determined by the generator

$$\boldsymbol{C}^\varepsilon(x)\varphi(u) = C^\varepsilon(u,x)\varphi'(u), \quad [1.49]$$

where

$$C^\varepsilon(u,x) = \varepsilon^{-1} C_1(u,x) + C_0(u,x),$$

and we also consider the operator $\boldsymbol{C}_\varepsilon(x)$ that has the form

$$\boldsymbol{C}_\varepsilon(x) := \varepsilon \boldsymbol{C}^\varepsilon(x) = \boldsymbol{C}_1(x) + \varepsilon \boldsymbol{C}_0(x). \quad [1.50]$$

Its components are determined by the following formulas:

$$\boldsymbol{C}_1(x)\varphi(u) := C_1(u,x)\varphi'(u), \boldsymbol{C}_0(x)\varphi(u) = C_0(u,x)\varphi'(u).$$

LEMMA 1.7.– *The compensating operator [1.48] of the EMRP [1.47] on test functions* $\varphi(u,x)$ *has the form:*

$$\boldsymbol{L}^{\varepsilon}\varphi(u,x) = \varepsilon^{-2}q(x)[\int_0^{\infty} G_x(ds)\boldsymbol{C}^{\varepsilon}_{\varepsilon^2 s}(x)\int_X P(x,dy)\varphi(u,y) - \varphi(u,x)]. \quad [1.51]$$

PROOF.– Since

$$E\varphi(u_1^{\varepsilon},x_1^{\varepsilon}) = E\boldsymbol{C}^{\varepsilon}_{\theta_x}(x)\varphi(u,x_1^{\varepsilon}) = \int_0^{\infty} G_x(ds)\boldsymbol{C}^{\varepsilon}_{\varepsilon^2 s}(x)\int_X P(x,dy)\varphi(u,y),$$

we obtain [1.51] from [1.48].

LEMMA 1.8.– *The compensating operator [1.51] on test functions* $\varphi(u,\cdot) \in C^3(R^d)$ *can be asymptotically represented by the relations:*

$$\begin{aligned}\boldsymbol{L}^{\varepsilon}\varphi(u,x) &= \varepsilon^{-2}Q\varphi(u,x) + \varepsilon^{-1}\theta_0^{\varepsilon}(x)\varphi(u,x) = \\ &= \varepsilon^{-2}Q\varphi(u,x) + \varepsilon^{-1}Q_1(x)\varphi(u,x) + \theta_1^{\varepsilon}(x)\varphi(u,x) = \\ &= \varepsilon^{-2}Q\varphi(u,x) + \varepsilon^{-1}Q_1(x)\varphi(u,x) + Q_2(x)\varphi(u,x) + \varepsilon\theta_2^{\varepsilon}(x)\varphi(u,x), \quad [1.52]\end{aligned}$$

where $Q_1(x)$ *and* $Q_2(x)$ *are operators that have the representations*

$$Q_1(x)\varphi(u,x) = \boldsymbol{C}_1(x)\boldsymbol{P}\varphi(u,x),$$

$$Q_2(x)\varphi(u,x) = [\,\boldsymbol{C}_0(x) + \mu_2(x)\boldsymbol{C}_1^2(x)]\boldsymbol{P}\varphi(u,x),$$

$$\mu_2(x) = \frac{g_2(x)}{2g(x)},$$

and the remainder terms in [1.52] has the form

$$\theta_0^{\varepsilon}(x)\varphi(u,x) = q(x)\boldsymbol{C}_{\varepsilon}(x)\boldsymbol{G}_1^{\varepsilon}(x)\,\boldsymbol{P}\varphi(u,x), \quad [1.53]$$

$$\theta_1^{\varepsilon}(x)\varphi(u,x) = q(x)[\boldsymbol{C}_{\varepsilon}^2(x)\boldsymbol{G}_2^{\varepsilon}(x) + \boldsymbol{C}_0(x)]\,\boldsymbol{P}\varphi(u,x), \quad [1.54]$$

$$\theta_2^{\varepsilon}(x)\varphi(u,x) = q(x)[\boldsymbol{C}_{\varepsilon}^3(x)\,\boldsymbol{G}_3^{\varepsilon}(x) + \frac{m_2(x)}{2}\boldsymbol{C}_2^{\varepsilon}(x)]\boldsymbol{P}\varphi(u,x). \quad [1.55]$$

The operators $\boldsymbol{G}_k^{\varepsilon}(x)$, $k = \overline{1,3}$, *are determined by the recursion equation*

$$\boldsymbol{G}_k^{\varepsilon}(x) = \int_0^{\infty} \overline{G}_x^{(k)}(s)ds\boldsymbol{C}^{\varepsilon}_{\varepsilon^2 s}(x),$$

where $\overline{G}_x^{(1)}(s) = \overline{G}_x(s)$, $\boldsymbol{C}_2^{\varepsilon}(x) = \boldsymbol{C}_0(x)[2\boldsymbol{C}_1(x) + \varepsilon\boldsymbol{C}_0(x)]$.

PROOF.– First, we use the obvious representation

$$\boldsymbol{L}^{\varepsilon} = \varepsilon^{-2}Q + \varepsilon^{-2}\boldsymbol{L}_1^{\varepsilon}(x), \qquad [1.56]$$

where

$$\boldsymbol{L}_1^{\varepsilon}(x) := q(x)[\boldsymbol{G}_{\varepsilon}(x) - I]\boldsymbol{P}, \qquad [1.57]$$

and, on the right side of [1.57], we obtain

$$\boldsymbol{G}_{\varepsilon}(x) := \int_0^{\infty} G_x(ds)\boldsymbol{C}_{\varepsilon^2 s}^{\varepsilon}(x). \qquad [1.58]$$

Furthermore, integrating by parts and equation for the semigroup

$$d\boldsymbol{C}_{\varepsilon^2 s}^{\varepsilon}(x) = \varepsilon^2 \boldsymbol{C}^{\varepsilon}(x)\boldsymbol{C}_{\varepsilon^2 s}^{\varepsilon}(x)ds,$$

and using condition $C3$ of the theorem for [1.58] give the following representation:

$$\boldsymbol{G}_{\varepsilon}(x) - I = \varepsilon^2\, \boldsymbol{C}^{\varepsilon}(x)\boldsymbol{G}_1^{\varepsilon}(x), \qquad [1.59]$$

where

$$\boldsymbol{G}_1^{\varepsilon}(x) = \int_0^{\infty} \overline{G}_x(s)dsC_{\varepsilon^2 s}^{\varepsilon}(x). \qquad [1.60]$$

Similar to [1.60], we obtain

$$\boldsymbol{G}_1^{\varepsilon}(x) = g(x)I + \varepsilon^2\boldsymbol{C}^{\varepsilon}(x)\boldsymbol{G}_2^{\varepsilon}(x), \qquad [1.61]$$

where

$$\boldsymbol{G}_2^{\varepsilon}(x) := \int_0^{\infty} \overline{G}_x^{(2)}(s)dsC_{\varepsilon^2 s}^{\varepsilon}(x), \qquad [1.62]$$

and from [1.62], we have

$$\boldsymbol{G}_2^{\varepsilon}(x) = \frac{g_2(x)}{2}I + \varepsilon^2\boldsymbol{C}^{\varepsilon}(x)\boldsymbol{G}_3^{\varepsilon}(x), \qquad [1.63]$$

where

$$\boldsymbol{G}_3^{\varepsilon}(x) := \int_0^{\infty} \overline{G}_x^{(3)}(s)dsC_{\varepsilon^2 s}^{\varepsilon}(x),$$

$$\overline{G}_x^{(3)}(s) := \int_s^{\infty} \overline{G}_x^{(2)}(s)ds.$$

By combining [1.60], [1.62] and [1.63], we obtain

$$\boldsymbol{G}_\varepsilon(x) - I = \varepsilon^2 g(x)\boldsymbol{C}^\varepsilon(x) + \varepsilon^4 g_2(x)[\boldsymbol{C}^\varepsilon(x)]^2 + \varepsilon^6[\boldsymbol{C}^\varepsilon(x)]^3\boldsymbol{G}_3^\varepsilon(x). \qquad [1.64]$$

From [1.49], [1.50] and [1.64], we now have

$$\boldsymbol{G}_\varepsilon(x) - I = \varepsilon g(x)\boldsymbol{C}_1(x) + \varepsilon^2[g(x)\boldsymbol{C}_0(x) + g_2(x)\boldsymbol{C}_1^2(x)] + \varepsilon^3\theta_3^\varepsilon(x), \qquad [1.65]$$

where the remainder term has the representation

$$\theta_3^\varepsilon(x) = g_2(x)(2\boldsymbol{C}_1(x)\boldsymbol{C}_0(x) + \varepsilon\boldsymbol{C}_0^2(x)) + \boldsymbol{C}_\varepsilon^3(x)\boldsymbol{G}_3^\varepsilon(x). \qquad [1.66]$$

By combining [1.57], [1.58], [1.65] and [1.66], we obtain all of the three asymptotic representations of [1.52].

Proof of Theorem 1.2 is based on using the solution of the singular perturbation problem for the compensating operator, which is given in the last asymptotic representation [1.52].

Let us consider the perturbed Lyapunov function:

$$V^\varepsilon(u,x) = V(u) + \varepsilon V_1(u,x) + \varepsilon^2 V_2(u,x), \qquad [1.67]$$

where $V(u)$ is a Lyapunov function for the limiting variance [1.41].

LEMMA 1.9.– *The compensating operator [1.51] has the representation*

$$\boldsymbol{L}^\varepsilon V^\varepsilon(u,x) = \boldsymbol{L}V(u) + \varepsilon\theta_L^\varepsilon(x)V(u)$$

on the perturbed Lyapunov function [1.67].

Here, $\boldsymbol{L}$ *is a generator of the limiting variance [1.41], and the remainder operator* $\theta_L^\varepsilon(x)$ *is determined by the relation*

$$\theta_L^\varepsilon(x)V(u) = \theta_2^\varepsilon(x)V(u) + \theta_1^\varepsilon(x)\boldsymbol{P}V_1(u,x) + \theta_0^\varepsilon(x)\boldsymbol{P}V_2(u,x). \qquad [1.68]$$

Perturbed Lyapunov functions have the representation

$$V_1(u,x) = R_0 Q_1(x)V(u), \qquad [1.69]$$

$$V_2(u,x) = R_0\widetilde{\boldsymbol{L}}(x)V(u), \qquad [1.70]$$

where

$$\widetilde{\boldsymbol{L}}(x) := \boldsymbol{L}(x) - \boldsymbol{L}, \qquad [1.71]$$

$$\boldsymbol{L}(x) := Q_2(x) + Q_1(x)\boldsymbol{P}R_0Q_1(x).$$

PROOF.– We consider the truncated operator for the compensating operator [1.52], i.e.

$$\boldsymbol{L}_0^\varepsilon = \varepsilon^{-2}Q + \varepsilon^{-1}Q_1(x) + Q_2(x).$$

For the operator $\boldsymbol{L}_0^\varepsilon$ on functions [1.67], we obtain the following representation:

$$\boldsymbol{L}_0^\varepsilon V^\varepsilon(u,x) = \varepsilon^{-2}QV(u) + \varepsilon^{-1}[QV_1(u,x) + Q_1(x)V(u)] \\ + [QV_2(u,x) + Q_1(x)V_1(u,x) + Q_2(x)V(u)] + \varepsilon\theta_{L0}^\varepsilon(x)V(u).$$

The SSPP for such decomposition gives [1.69] and [1.70] of the perturbed Lyapunov function [1.67], and the limiting operator $\boldsymbol{L}$, that can be calculated by the formula

$$\boldsymbol{L}\Pi = \Pi Q_2(x)\Pi + \Pi Q_1(x)R_0Q_1(x)\Pi.$$

Let us denote each of the representations of the compensating operator [1.52] as

$$\boldsymbol{L}_{(0)}^\varepsilon = \varepsilon^{-2}Q + \varepsilon^{-1}\theta_0^\varepsilon(x),$$

$$\boldsymbol{L}_{(1)}^\varepsilon = \varepsilon^{-2}Q + \varepsilon^{-1}Q_1(x) + \theta_1^\varepsilon(x),$$

$$\boldsymbol{L}_{(2)}^\varepsilon = \varepsilon^{-2}Q + \varepsilon^{-1}Q_1(x) + Q_2(x) + \varepsilon\theta_2^\varepsilon(x).$$

Then, the compensating operator $\boldsymbol{L}^\varepsilon$ on the perturbed Lyapunov function $V^\varepsilon(u,x)$ has the decomposition

$$\boldsymbol{L}^\varepsilon V^\varepsilon(u,x) = \varepsilon^2\boldsymbol{L}_{(0)}^\varepsilon V_2(u,x) + \varepsilon\boldsymbol{L}_{(1)}^\varepsilon V_1(u,x) + \boldsymbol{L}_{(2)}^\varepsilon V(u).$$

Since

$$\varepsilon^2\boldsymbol{L}_{(0)}^\varepsilon V_2(u,x) = QV_2(u,x) + \varepsilon\theta_0^\varepsilon(x)V_2(u,x),$$

$$\varepsilon\boldsymbol{L}_{(1)}^\varepsilon V_1(u,x) = \varepsilon^{-1}QV_1(u,x) + Q_1(x)V_1(u,x) + \varepsilon\theta_1^\varepsilon(x)V_1(u,x),$$

$$\boldsymbol{L}_{(2)}^\varepsilon V(u) = \varepsilon^{-2}QV(u) + \varepsilon^{-1}Q_1(x)V(u) + Q_2(x)V(u) + \varepsilon\theta_2^\varepsilon(x)V(u),$$

then

$$\boldsymbol{L}^\varepsilon V^\varepsilon(u,x) = \varepsilon^{-2}QV(u) + \varepsilon^{-1}[QV_1(u,x) + Q_1(x)V(u)] \\ + [QV_2(u,x) + Q_1(x)V_1(u,x) + Q_2(x)V(u)] + \varepsilon\theta_L^\varepsilon(x)V(u),$$

where the remainder operator $\theta_L^\varepsilon(x)$ has the representation as in [1.68]. The proof of the lemma is completed.

Thus, the limiting operator of the averaged variance [1.41] is determined by equality [1.45]:

$$\boldsymbol{L}\varphi(u) = C(u)\varphi'(u) + \frac{1}{2}Sp[B(u)\varphi''(u)],$$

where $C(u)$ and $B(u)$ are calculated by formulas [1.46] and [1.43] respectively.

Using the representations [1.69] and [1.70] of the perturbed functions

$$V_k(u, x), \ k = 1, 2,$$

and the term [1.68] for the remainder operator $\theta_L^\varepsilon(x)$, we obtain the following representation of the remainder operator on Lyapunov functions $V(u)$:

$$\theta_L^\varepsilon(x)V(u) = [\theta_2^\varepsilon(x) + \theta_1^\varepsilon(x)\boldsymbol{P}R_0Q_1(x) + \theta_0^\varepsilon(x)\boldsymbol{P}R_0\tilde{\boldsymbol{L}}(x)]V(u).$$

From the relations [1.53]–[1.55] for the remainder operators

$$\theta_k^\varepsilon(x), \ k = 0, 1, 2,$$

and the representation [1.71] $\widetilde{\boldsymbol{L}}(x)$, we conclude that, in the remainder operator $\theta_L^\varepsilon(x)$, differentiation operators act on the variable $u \in R^d$ of not higher than the third order.

Moreover, from the conditions of the theorem, it follows that the operators $\boldsymbol{G}_k^\varepsilon(x), k = 1, 2, 3$, and the potential R_0 are bounded on the functional space $V(u) \in C^3(R^d)$.

We note that, under the conditions of Theorem 1.2, the estimation

$$|\theta_L^\varepsilon(x)V(u)| \le c_L V(u)$$

holds true.

Under the conditions of Theorem 1.2, $C1$–$C3$ for all $\varepsilon \le \varepsilon_0$, where ε_0 is small enough ($\varepsilon_0 \le c/c_L$), holds true for the base inequality:

$$\boldsymbol{L}^\varepsilon V^\varepsilon(u, x) \le -cV(u), c > 0. \qquad [1.72]$$

PROOF OF THEOREM 1.2.– The proof of Theorem 1.2 is the same as the proof of Theorem 1.1.

The representation [1.69] and [1.70] of the perturbed functions $V_k(u, x), k = 1, 2$, and conditions $C2$ of the theorem give the following estimation for the perturbed Lyapunov function $V^\varepsilon(u, x)$:

$$0 < (1 - \varepsilon c)V(u) \le V^\varepsilon(u, x) \le (1 + \varepsilon c)V(u). \qquad [1.73]$$

The martingale characterization of the process $\eta^\varepsilon(t) := V^\varepsilon(u^\varepsilon(\tau^\varepsilon(t)), x(t/\varepsilon^2))$, where $\tau^\varepsilon(t) = \tau_{\nu(t)}^\varepsilon$, and the base inequality [1.72] describe the process $\eta^\varepsilon(t)$ as non-negative supermartingale. Thus, a non-negative limit v^ε exists:

$$P\{\lim_{t\to\infty} V^\varepsilon(u^\varepsilon(\tau^\varepsilon(t)), x(t/\varepsilon^2)) = v^\varepsilon\} = 1$$

with probability 1.

At the same time, the random variable v^ε has a limited expectation, since

$$EV^\varepsilon(u^\varepsilon(\tau^\varepsilon(t)), x(t/\varepsilon^2)) \leq V^\varepsilon(u, x) \leq (1 + \varepsilon c)V(u).$$

From the additional property of the Lyapunov function

$$V(u) \to \infty, u \to \infty, \qquad [1.74]$$

we conclude that $P\{v^\varepsilon < \infty\} = 1$. Again, the base inequality [1.72] and the estimate [1.73] yield

$$P\{\lim_{t\to\infty} V(u^\varepsilon(\tau^\varepsilon(t)) = 0\} = 1,$$

i.e.

$$P\{\lim_{n\to\infty} u^\varepsilon(\tau_n) = 0\} = 1.$$

Finally, the positivity of the Lyapunov function $V(u) > 0$, at $u \neq 0$, the property [1.74] and the regularity of the semi-Markov process $x(t), t \geq 0$, complete the proof of the theorem.

2

Levy Approximation Scheme

2.1. Differential equations with small stochastic additions in the Levy approximation scheme

In this section, we consider the case where the disturbances of the system are defined by an impulse process in the nonclassical Levy approximation scheme (Korolyuk and Limnios 2005), which will play an important role in determining the asymptotic behavior of the generator of this system.

The stochastic evolutionary system in an ergodic Markov medium is defined by the stochastic differential equation

$$du^{\varepsilon}(t) = C(u^{\varepsilon}, x(t/\varepsilon^2))dt + d\eta^{\varepsilon}(t), \ u^{\varepsilon}(t) \in \mathbf{R}. \qquad [2.1]$$

where a uniformly ergodic Markov process $x(t)$ in a standard phase space $(X, \mathbf{X})$ is defined by the generator

$$\mathbf{Q}\varphi(x) = q(x) \int_X P(x, dy)[\varphi(y) - \varphi(x)]$$

on the Banach space $B(X)$ of real-valued bounded functions $\varphi(x)$ with the supremum norm $\|\varphi\| = \max_{x \in X} |\varphi(x)|$.

The stochastic kernel $P(x, B)$, $x \in X$, $B \in \mathbf{X}$ defines a uniformly ergodic embedded Markov chain $x_n = x(\tau_n)$, $n \geq 0$, with the stationary distribution $\rho(B)$, $B \in \mathbf{X}$. The stationary distribution $\pi(B)$, $B \in \mathbf{X}$ of the Markov process $x(t)$, $t \geq 0$, is defined by the relation

$$\pi(dx)q(x) = q\rho(dx), \ q = \int_X \pi(dx)q(x).$$

We denote the potential operator of the generator $\mathbf{Q}$ by R_0, which is defined by the equality Korolyuk and Limnios (2005)

$$R_0 = \Pi - (\Pi + \mathbf{Q})^{-1},$$

where $\Pi\varphi(x) = \int\limits_X \pi(dy)\varphi(y)1(x)$ is the projector of zeros of the generator $\mathbf{Q}$ onto the subspace $N_Q = \{\varphi : \ \mathbf{Q}\varphi = 0\}$.

The impulse perturbation process (IPP) $\eta^\varepsilon(t)$, $t \geq 0$, in the Levy approximation scheme is defined by the relation

$$\eta^\varepsilon(t) = \int\limits_0^t \eta^\varepsilon(ds, x(s/\varepsilon^2)), \qquad [2.2]$$

where the family of processes with independent increments $\eta^\varepsilon(t, x)$, $t \geq 0$, $x \in X$, is defined by the generators

$$\Gamma^\epsilon(x)\varphi(w) = \varepsilon^{-2} \int\limits_R (\varphi(w + v) - \varphi(w))\Gamma^\varepsilon(dv, x), \ \ x \in X \qquad [2.3]$$

and satisfies the properties of the Levy approximation Korolyuk and Limnios (2005).

L1. The approximation of averages

$$\int\limits_R v\Gamma^\varepsilon(dv, x) = \varepsilon a_1(x) + \varepsilon^2(a_2(x) + \theta_a(x)), \ \theta_a(x) \to 0, \ \varepsilon \to 0,$$

and

$$\int\limits_R v^2\Gamma^\varepsilon(dv, x) = \varepsilon(b(x) + \theta_b(x)), \ \theta_b(x) \to 0, \ \varepsilon \to 0,$$

L2. The condition imposed on the distribution function

$$\int\limits_R g(v)\Gamma^\varepsilon(dv, x) = \varepsilon^2(\Gamma_g(x) + \theta_g(x)), \ \theta_g(x) \to 0, \ \varepsilon \to 0,$$

for all $g(v) \in C^2(\mathbb{R})$ (the space of real-valued bounded functions such that $g(v)/|v|^2 \to 0$, $|v| \to 0$), where the measure $\Gamma_g(x)$ is bounded for all $g(v) \in C^2(\mathbb{R})$ and defined by the relation (functions from the space $C^2(\mathbb{R})$ separate the measures)

$$\Gamma_g(x) = \int\limits_R g(v)\Gamma_0(dv, x), \ g(v) \in C^3(\mathbb{R});$$

L3. The uniform quadratic integrability

$$\sup \lim_{c\to\infty} \int_{|v|>c} v^2 \Gamma_0(dv, x) = 0;$$

Let us assume that the balance condition is fulfilled:

$$\hat{a}_1 := \int_X \pi(dx) a_1(x) = 0. \qquad [2.4]$$

Consider the asymptotic properties of the perturbation process.

THEOREM 2.1.– *If the balance condition [2.4] and conditions L1 – L3 are satisfied, the following weak convergence*

$$\eta^\varepsilon(t) \to \eta^0(t), \ \varepsilon \to 0$$

takes place for IPP [2.3].

The limiting process $\eta^0(t)$ is defined by the generator

$$\Gamma\varphi(w) = \hat{a}_2\varphi'(w) + \frac{1}{2}\sigma^2\varphi''(w) + \int_R [\varphi(w+v) - \varphi(w)]\hat{\Gamma}_0(dv),$$

where

$$\hat{a} = \int_X \pi(dx)(a_2(x) - a_0(x)),$$

$$\sigma^2 = \int_X \pi(dx)(b(x) - b_0(x)) + 2\int_X \pi(dx) a_1(x) R_0 a_1(x),$$

$$a_0(x) = \int_R v\Gamma_0(dv, x),$$

$$b_0(x) = \int_R v^2\Gamma_0(dv, x),$$

$$\hat{\Gamma}_0(v) = \int_X \pi(dx)\Gamma_0(v, x),$$

and is a Levy process, which has three components: deterministic drift, diffusion component and Poisson jump component.

PROOF.– First, Let us consider some propositions.

LEMMA 2.1.– *Generators of processes with independent increments* $\eta^\varepsilon(t,x)$, $t \geq 0$, $x \in X$, *on test functions* $\varphi(w) \in C^2(\mathbf{R})$ *under conditions L1 – L3 have the asymptotic representation*

$$\mathbf{\Gamma}^\varepsilon(x)\varphi(w) = \varepsilon^{-1}\Gamma_1(x)\varphi(w) + \Gamma_2(x)\varphi(w), \tag{2.5}$$

where

$$\Gamma_1(x)\varphi(w) = a_1(x)\varphi'(w),$$

$$\Gamma_2(x)\varphi(w) = (a_2(x) - a_0(x))\varphi'(x) + \frac{1}{2}(b(x) - b_0(x))\varphi''(x) +$$
$$+ \int_R [\varphi(w+v) - \varphi(v)]\Gamma_0(dv, x).$$

PROOF.– Using the expansion of the function $\varphi(w)$ into the Taylor series, we transform the generator [2.3]:

$$\begin{aligned}
\mathbf{\Gamma}^\varepsilon(x)\varphi(w) &= \varepsilon^{-2}\int_R (\varphi(w+v) - \varphi(v))\Gamma^\varepsilon(dv, x) \\
&= \varepsilon^{-2}\int_R (\varphi(w+v) - \varphi(v) - v\varphi'(v) - \frac{1}{2}v^2\varphi''(w))\Gamma^\varepsilon(dv, x) \\
&\quad + \varepsilon^{-2}\int_R (v\varphi'(w)\Gamma^\varepsilon(dv, x) + \frac{1}{2}v^2\varepsilon^{-2}\int_R v^2\varphi''(w)\Gamma^\varepsilon(dv, x) \\
&= \int_R (\varphi(u+v) - \varphi(v) - v\varphi'(w) - \frac{1}{2}v^2\varphi''(w))\Gamma_0(dv, x) \\
&\quad + \varepsilon^{-1}a_1(x)\varphi'(w) + a_2(x)\varphi'(w) + \frac{1}{2}(b(x) - b_0(x))\varphi''(w) \\
&\quad + \int_R (\varphi(u+v) - \varphi(v))\Gamma_0(dv, x) + \gamma^\varepsilon(w)\varphi(w),
\end{aligned}$$

where the penultimate equality follows from conditions L1 and L2. Note that the function $\varphi(w+v) - \varphi(w) - v\varphi'(w) - \frac{1}{2}v^2\varphi''(w) \in C^2(\mathbf{R})$, since it is bounded due to the boundedness of $\varphi(w)$ and its derivatives and

$$[\varphi(w+v) - \varphi(w) - v\varphi'(w) - \frac{1}{2}v^2\varphi''(w)]/|v^2| \to 0$$

when $v \to 0$.

Taking into account that $\gamma^\varepsilon(w)\varphi(w) = o(\varepsilon^2)$, $\varphi(w) \in C^2(\mathbf{R})$, we obtain representation [2.5].

LEMMA 2.2.– *The generator of a two-component Markov process* $(\eta^\varepsilon, x(t/\varepsilon^2))$, $t \geq 0$, *has the form*

$$\hat{\Gamma}^\varepsilon(x)\varphi(w,x) = \varepsilon^{-2}\mathbf{Q}\varphi(w,x) + \varepsilon^{-1}\Gamma_1(x)\varphi(w,x) + \Gamma_2(x)\varphi(w,x) + \gamma^\varepsilon(x)\varphi(w,x), \quad [2.6]$$

where operators $\Gamma_1(x)$ *and* $\Gamma_2(x)$ *are defined in Lemma 2.1 and the remainder term* $\|\gamma^\varepsilon(x)\varphi(w,x)\| \to 0$ *as* $\varepsilon \to 0$, $\varphi(w,\cdot) \in C^2(\mathbf{R})$.

PROOF.– The statement of Lemma 2 becomes obvious if we use the definition of the generator of the Markov process and the form of the respective generators of processes $\eta^\varepsilon(t,x)$ and $x(t/\varepsilon^2)$.

The truncated operator has the form Chabanyuk (2004b)

$$\Gamma_0^\varepsilon(x)\varphi(w) = \varepsilon^{-2}\mathbf{Q}\varphi(w,x) + \varepsilon^{-1}\Gamma_1(x)\varphi(w,x) + \Gamma_2(x)\varphi(w,x). \quad [2.7]$$

LEMMA 2.3.– *Under the balance condition [2.4], the solution of the singular perturbation problem for the truncated operator [2.7] on test functions*

$$\varphi^\varepsilon(w,x) = \varphi(w) + \varepsilon\varphi_1(w,x) + \varepsilon^2\varphi_2(w,x)$$

is implemented by the relation

$$\Gamma_0^\varepsilon(x)\varphi^\varepsilon(w,x) = \Gamma\varphi(w) + \varepsilon\theta_\eta^\varepsilon(x)\varphi(w), \quad [2.8]$$

where the remainder term $\theta_\eta^\varepsilon(x)\varphi(w)$ *is uniformly bounded in* x.

The limiting operator is defined by the formula

$$\Gamma = \Pi\Gamma_1(X)R_0\Gamma_1(x)\Pi + \Pi\Gamma_2(x)\Pi. \quad [2.9]$$

PROOF.– For equality [2.8] to hold, the coefficients of identical degrees ε on the left and the right should coincide. Let us calculate

$$\begin{aligned}\Gamma_0^\varepsilon(x)\varphi^\varepsilon(w,x) &= \varepsilon^{-2}\mathbf{Q}\varphi(w)\\ &+ \varepsilon^{-1}[\mathbf{Q}\varphi_1(w,x) + \Gamma_1(x)\varphi(w)]\\ &+ [\mathbf{Q}\varphi_2(w,x) + \Gamma_1(x)\varphi_1(w,x) + \Gamma_2(x)\varphi_2(w,x)]\\ &+ \varepsilon[\Gamma_1(x)\varphi_2(w,x) + \Gamma_2(x)\varphi_1(w,x)] + \varepsilon^2\Gamma_2(x)\varphi_2(w,x).\end{aligned}$$

From the first summand, we obtain

$$\mathbf{Q}\varphi(w) = 0 \Leftrightarrow \varphi(w) \in N_{\mathbf{Q}}.$$

From the above equation, we can see that $\varphi(w)$ does not depend on x.

The balance condition [2.4] is a condition of solvability of the equation

$$\mathbf{Q}\varphi_1(w,x) + \Gamma_1(x)\varphi(w) = 0.$$

Then

$$\varphi_1(w,x) = R_0\Gamma_1(x)\varphi(w). \qquad [2.10]$$

The equation

$$\mathbf{Q}\varphi_2(w,x) + \Gamma_1(x)\varphi_1(w,x) + \Gamma_2(x)\varphi(w) = \Gamma\varphi(w)$$

according to [2.10] can be rewritten as

$$\mathbf{Q}\varphi_2(w,x) + \Gamma_1(x)R_0\Gamma_1(x)\varphi(w,x) + \Gamma_2(x)\varphi(w) = \Gamma\varphi(w).$$

The resolvability condition for the last equation gives the limiting operator in the form [2.9]. Then

$$\varphi_2(w,x) = R_0[\Gamma_1(x)R_0\Gamma_1(x) + \Gamma_2(x) - \Gamma]\varphi(w). \qquad [2.11]$$

Using [2.10] and [2.11], we can reduce the other terms of the expansion to the form

$$\begin{aligned}
&\varepsilon[\Gamma_1(x)\varphi_2(w,x) + \Gamma_2(x)\varphi_1(w,x)] + \varepsilon^2\Gamma_2(x)\varphi(w,x) \\
&= \varepsilon[[\Gamma_1(x)R_0[\Gamma_1(x)R_0\Gamma_1(x) + \Gamma_2(x) - \Gamma] + \Gamma_2(x)R_0\Gamma_1(x)] \\
&\quad + \varepsilon\Gamma_2(x)R_0[\Gamma_1(x)R_0\Gamma_1(x) + \Gamma_2(x) - \Gamma]]\varphi(w) = \varepsilon\theta_\eta^\varepsilon(x)\varphi(w).
\end{aligned}$$

The boundedness of $\theta_\eta^\varepsilon(x)\varphi(w)$ follows from the form of operators Γ_1, Γ_2 and R_0.

The proof of Lemma 2.3 is completed.

The proof of Theorem 2.1 is completed by means of Lemma 2.3 and Theorem 4.2 from Korolyuk and Limnios (2005).

Consider the asymptotic properties of the original evolutionary system [2.1].

THEOREM 2.2.– *If the balance condition and conditions L1 – L3 are satisfied, the following weak convergence*

$$u^\varepsilon(t) \to \hat{u}(t), \ \ \varepsilon \to 0.$$

takes place.

The limiting process $\hat{u}(t)$ is defined by the generator

$$\mathbf{L}\varphi(w) = \hat{C}(u)\varphi'(w) + \Gamma\varphi(w), \qquad [2.12]$$

where $\hat{C}(u) = \int\limits_X \pi(dx)C(u,x)$.

NOTE 2.1.– *The weak convergence of processes* $u^\varepsilon(t) \to \hat{u}(t)$, $\varepsilon \to 0$, *follows from the convergence of the respective generators when the compactness of the prelimiting set of processes* $u^\varepsilon(t)$ *holds true.*

NOTE 2.2.– *The limiting process* $\hat{u}(t)$ *can be given by the stochastic differential equation*

$$d\hat{u}_d(t) = [\hat{C}(\hat{u}(t)) + \hat{a}_2]dt + \sigma dW(t) + \int_{\mathbb{R}} v\tilde{v}(dt, dv),$$

where $\mathbf{E}\tilde{\nu}(dt, dv) = dt\tilde{\Gamma}_0(dv)$.

NOTE 2.3.– *The limiting process* $\hat{u}(t)$ *has three components. The deterministic drift is defined by the solution of the differential equation*

$$d\hat{u}_d(t) = [\hat{C}(\hat{u}_d(t)) + \hat{a}_2]dt, \qquad [2.13]$$

where the additional term $\hat{a}_2$ *appears due to accumulation with the normalized time* t/ε^2, $\varepsilon \to 0$ *of small jumps of the impulse process that occur with probability close to 1. The second, diffusion component, is defined by a parameter* σ, *which arises due to accumulation with growth of the normalized time* t/ε^2, $\varepsilon \to 0$ *of small jumps of degree* ε, *which also occur with probability close to 1.*

The third component is rare big jumps that take place with nearly zero probability and are defined in terms of averaged measure of jumps $\tilde{\Gamma}_0(dv)$ *by the generator*

$$\Gamma_j\varphi(w) = \int_R [\varphi(w + v) - \varphi(w)]\tilde{\Gamma}_0(dv).$$

PROOF OF THEOREM 2.2.–

LEMMA 2.4.– *The generator of a two-component Markov process* $(u^\varepsilon(t), x(t/\varepsilon^2))$, $t \geq 0$, *has the representation*

$$\begin{aligned}\mathbf{L}^\varepsilon(x)\varphi(w,x) &= \varepsilon^{-2}\mathbf{Q}\varphi(w,x) + \Gamma^\varepsilon(x)\varphi(w,x)\\ &\quad + \mathbf{C}(x)\varphi(w,x) + \theta^\varepsilon_w\varphi(w,x),\end{aligned} \qquad [2.14]$$

where $\Gamma^\varepsilon(x)$ *is the generator of the family of processes with independent increments [2.3]:*

$$\mathbf{C}(x)\varphi(w,x) = C(u,x)\varphi'_w(w,x).$$

The remainder term $\|\theta^\varepsilon_w\varphi(w,x)\| \to 0$ *as* $\varepsilon \to 0$.

The proof of a similar lemma is presented in Chabanyuk (2007).

LEMMA 2.5.– *The generator* $\mathbf{L}^\varepsilon(x)$ *has the asymptotic representation*

$$\mathbf{L}^\varepsilon(x)\varphi(w,x) = \varepsilon^{-2}\mathbf{Q}\varphi(w,x) + \varepsilon^{-1}\Gamma_1(x)\varphi(w,x) + \Gamma_2(x)\varphi(w,x) + \mathbf{C}(x)\varphi(w,x) + \hat{\theta}^\varepsilon_w\varphi(w,x), \qquad [2.15]$$

where

$$\hat{\theta}^\varepsilon_w(x) = \gamma^\varepsilon + \theta^\varepsilon_w(x),$$

$\Gamma_1(x)$ $\Gamma_2(x)$ *1.*

The remainder term

$$\|\hat{\theta}^\varepsilon_w\varphi(w,x)\| \to 0$$

when $\varepsilon \to 0$.

The proof uses the representation of the operator [2.5] and the results of Lemma 2.4.

The truncated operator has the form

$$\mathbf{L}^\varepsilon_0(x)\varphi = \varepsilon^2\mathbf{Q}\varphi + \varepsilon^{-1}\Gamma_1(x)\varphi + \Gamma_2(x)\varphi + \mathbf{C}(x)\varphi. \qquad [2.16]$$

LEMMA 2.6.– *Under the balance condition [2.4], the solution of the singular perturbation problem for the truncated operator [2.16] on test functions*

$$\varphi^\varepsilon(w,x) = \varphi(w) + \varepsilon\varphi_1(w,x) + \varepsilon^2\varphi(w,x)$$

is carried out by the relation

$$\mathbf{L}^\varepsilon_0(x)\varphi^\varepsilon(w,x) = \mathbf{L}\varphi(w) + \varepsilon^2\theta^\varepsilon_w(x)\varphi(w), \qquad [2.17]$$

where the remainder term $\theta^\varepsilon_w(x)$ *is uniformly bounded with respect to* x

The limiting operator $\mathbf{L}$ *is defined by the formula*

$$\mathbf{L} = \Pi[\mathbf{C}(x) + \Gamma_1(x)R_0\Gamma_1(x) + \Gamma_2(x)]\Pi. \qquad [2.18]$$

PROOF.– For equality [2.17] to hold, the coefficients of identical degrees ε on the left and the right should coincide. Thus, we will calculate

$$\begin{aligned}\mathbf{L}^\varepsilon_0(x)\varphi^\varepsilon(w,x) &= \varepsilon^{-2}\mathbf{Q}(x)\varphi(w) \\ &+ \varepsilon^{-1}[\mathbf{Q}\varphi_1(w,x) + \Gamma_1(x)\varphi(w)] \\ &+ [\mathbf{Q}\varphi_2(w,x) + \Gamma_1(x)\varphi_1(w,x)] \\ &+ \varepsilon[\Gamma_1(x)\varphi_2(w,x) + \Gamma_2(x)\varphi_1(w,x)) + \mathbf{C}(x)\varphi_1(w,x)] \\ &+ \varepsilon^2[\Gamma_2(x)\varphi_2(w,x) + \mathbf{C}(x)\varphi_2(w,x)].\end{aligned}$$

Since

$$\mathbf{Q}\varphi(w) \Leftrightarrow \varphi(w) \in N_Q,$$

it is obvious that $\varphi(w)$ does not depend on x.

The balance condition [2.4] is the condition of resolvability to the equation

$$\mathbf{Q}\varphi_1(w,x) + \Gamma_1(x)\varphi(w) = 0.$$

Then

$$\varphi_1(w,x) = R_0\Gamma_1(x)\varphi(w).$$

We have the last equation

$$\mathbf{Q}\varphi_2(w,x) + \Gamma_1(x)\varphi(w,x) + \Gamma_2(x)\varphi(w) + \mathbf{C}(x)\varphi(w) = \mathbf{L}\varphi(w).$$

This can be rewritten in the form

$$\mathbf{Q}\varphi_2(w,x) = [\mathbf{L} - \Gamma_1(x)R_0\Gamma_1(x) - \Gamma_2(x) - \mathbf{C}(x)]\varphi(w).$$

The condition of resolvability of the last equation gives the limiting operator $\mathbf{L}$ in the form [2.18].

The proof of the theorem is completed, which is similar to the proof of Theorem 4.2 from Korolyuk and Limnios (2005).

2.2. Asymptotic dissipativity of stochastic processes with impulse perturbations in the Levy approximation scheme

In this section, we investigate the following question: how does the behavior of the limiting process depend on the prelimiting normalization of the stochastic evolutionary system in an ergodic Markov environment.

We consider the evolutionary system [2.1] and suppose that the assumptions of the previous section hold true.

THEOREM 2.3.– *Let us assume that the Lyapunov function $V(u) \in C^3(\mathbb{R}^d)$ of the system exists:*

$$\frac{du}{dt} = \alpha(u), \qquad [2.19]$$

where $\alpha(u) = \hat{C}(u) + \hat{a}$, and the following conditions and inequalities are fulfilled

C1: $|\Gamma_u^1(x)R_0\hat{L}V(u)| < M_1V(u),\ M_1 > 0;$
C2: $|\Gamma_u^1(x)R_0\Gamma_u^1(x)V(u)| < M_2V(u),\ M_2 > 0;$
C3: $|\Gamma_u^1(x)R_0(C)(x)V(u)| < M_3V(u),\ M_3 > 0;$
C4: $|\mathbf{C}(x)R_0\hat{L}V(u)| < M_4V(u),\ M_4 > 0;$
C5: $|\mathbf{C}(x)R_0\Gamma_u^1(x)V(u)| < M_5V(u),\ M_5 > 0;$
C6: $|\mathbf{C}(x)R_0\mathbf{C}(x)V(u)| < M_6V(u),\ M_6 > 0.$

$$\alpha(u)V'(u) < -c_1V(u), \quad [2.20]$$

$$\sup_{u\in R^d} \|\sigma(u)\| < c_2, \quad [2.21]$$

$$|\int_X v^2\Gamma_0(dv,x)| < c_3(x), \quad [2.22]$$

where $c_1 > 0$, $c_2 > 0$,

$$\hat{c}_3 = \int_X \pi(dx)c_3(x) > 0.$$

Then, system [2.1] is asymptotic dissipativity.

First, let us prove some propositions.

LEMMA 2.7.– *The generator of a three-component Markov process* $u^\varepsilon(t), x(t/\varepsilon : 2, \eta^\varepsilon(t,x)), t \geq 0$, *can be represented in the form*

$$\mathbf{L}^\varepsilon(x)\varphi(u,w,x) = \varepsilon^{-1}\mathbf{Q}\varphi(u,w,x) + \Gamma_w^\varepsilon(x)\varphi(u,w,x) + \mathbf{C}(x)\varphi(u,w,x) + \Gamma_u^\varepsilon(x)\varphi(u,w,x), \quad [2.23]$$

where $\Gamma_w^\varepsilon(x)$ *is the generator of a set of processes with independent increments [2.3], acting at the variable* w, *and* $\Gamma_u^\varepsilon(x)$ *is the generator of a set of processes with independent increments [2.3], acting at the variable* u.

PROOF.– The generator of the Markov process on the perturbed test function is determined by the relation

$$\mathbf{L}^\varepsilon(x)\varphi(u,w,x) = \lim_{\Delta\to 0}\frac{1}{\Delta}\mathbb{E}[\varphi(u_{t+\Delta}^\varepsilon, w_{t+\Delta}^\varepsilon, x_{t+\Delta}^\varepsilon) - \varphi(u,w,x)|u^\varepsilon(t) = u,\ \eta^\varepsilon(t) = w,\ x(t/\varepsilon^2) = x].$$

Let us add and subtract the conditional expectation $\varphi(u, w^{\varepsilon}_{t+\Delta}, x^{\varepsilon}_{t+\Delta})$

$$\begin{aligned}&\mathbb{E}[\varphi(u^{\varepsilon}_{t+\Delta}, w^{\varepsilon}_{t+\Delta}, x^{\varepsilon}_{t+\Delta})\\&-\varphi(u,w,x)|u^{\varepsilon}(t)=u,\ \eta^{\varepsilon}(t)=w,\ x(t/\varepsilon^2)=x]\\&=\mathbb{E}[\varphi(u^{\varepsilon}_{t+\Delta}, w^{\varepsilon}_{t+\Delta}, x^{\varepsilon}_{t+\Delta})-\varphi(u, w^{\varepsilon}_{t+\Delta}, x^{\varepsilon}_{t+\Delta})]\\&=\mathbb{E}[\varphi(u,w^{\varepsilon}_{t+\Delta}, x^{\varepsilon}_{t+\Delta})-\varphi(u,w,x)].\end{aligned}$$

The expansion of $u^{\varepsilon}_{t+\Delta}$ has the form

$$u^{\varepsilon}_{t+\Delta}=u+C(u,x)\Delta+\Delta w+o(\Delta).$$

The expression obtained will be substituted in the first addition of the conditional expectation

$$\begin{aligned}&\mathbb{E}[\varphi(u^{\varepsilon}_{t+\Delta}, w^{\varepsilon}_{t+\Delta}, x^{\varepsilon}_{t+\Delta})-\varphi(u, w^{\varepsilon}_{t+\Delta}, x^{\varepsilon}_{t+\Delta})]\\&=\mathbb{E}[\varphi(u+C(u,x)\Delta+\Delta w+o(\Delta), w^{\varepsilon}_{t+\Delta}, x^{\varepsilon}_{t+\Delta})-\varphi(u, w^{\varepsilon}_{t+\Delta}, x^{\varepsilon}_{t+\Delta})]\\&=\mathbb{E}[\varphi(z+\Delta z, w^{\varepsilon}_{t+\Delta}, x^{\varepsilon}_{t+\Delta})-\varphi(u, w^{\varepsilon}_{t+\Delta}, x^{\varepsilon}_{t+\Delta})],\end{aligned}$$

where

$$z=u+C(u,x)\Delta+o(\Delta).$$

Then, we add and subtract $\varphi(z, w^{\varepsilon}_{t+\Delta}, x^{\varepsilon}_{t+\Delta})$ in the expression obtained

$$\begin{aligned}&\mathbb{E}[\varphi(u^{\varepsilon}_{t+\Delta}, w^{\varepsilon}_{t+\Delta}, x^{\varepsilon}_{t+\Delta})-\varphi(u, w^{\varepsilon}_{t+\Delta}, x^{\varepsilon}_{t+\Delta})]\\&=\mathbb{E}[\varphi(z+\Delta w, w^{\varepsilon}_{t+\Delta}, x^{\varepsilon}_{t+\Delta})-\varphi(z, w^{\varepsilon}_{t+\Delta}, x^{\varepsilon}_{t+\Delta})]\\&\quad+\mathbb{E}[\varphi(z, w^{\varepsilon}_{t+\Delta}, x^{\varepsilon}_{t+\Delta})-\varphi(u, w^{\varepsilon}_{t+\Delta}, x^{\varepsilon}_{t+\Delta})].\end{aligned}$$

Since the generator $\Gamma^{\varepsilon}_{u}(x)$ has the form

$$\Gamma^{\varepsilon}_{u}(x)\varphi(u,w,x)=\lim_{\Delta\to 0}\frac{1}{\Delta}(\mathbb{E}[\varphi(u+\Delta u,w,x)-\varphi(u,w,x)]),$$

for the limit of the first addition, we have

$$\lim_{\Delta\to 0}\frac{1}{\Delta}\mathbb{E}[\varphi(z+\Delta w, w^{\varepsilon}_{t+\Delta}, x^{\varepsilon}_{t+\Delta})-\varphi(z, w^{\varepsilon}_{t+\Delta}, x^{\varepsilon}_{t+\Delta})]=\Gamma^{\varepsilon}_{u}(x)\varphi(u,w,x)$$

Let us expand $\varphi(z, w^{\varepsilon}_{t+\Delta}, x^{\varepsilon}_{t+\Delta})$ according to the Taylor formula:

$$\begin{aligned}&\varphi(u+C(u,x)\Delta+o(\Delta), w^{\varepsilon}_{t+\Delta}, x^{\varepsilon}_{t+\Delta})\\&=\varphi(u, w^{\varepsilon}_{t+\Delta}, x^{\varepsilon}_{t+\Delta})+\varphi', w^{\varepsilon}_{t+\Delta}, x^{\varepsilon}_{t+\Delta})(C(u,x)\Delta+o(\Delta))+o(\Delta).\end{aligned}$$

Substituting the expansion obtained in the expression $\mathbb{E}[\varphi(z, w^{\varepsilon}_{t+\Delta}, x^{\varepsilon}_{t+\Delta}) - \varphi(u, w^{\varepsilon}_{t+\Delta}, x^{\varepsilon}_{t+\Delta})]$, we have

$$\begin{aligned}
&\lim_{\Delta\to 0}\frac{1}{\Delta}\mathbb{E}[\varphi(z, w^{\varepsilon}_{t+\Delta}, x^{\varepsilon}_{t+\Delta}) - \varphi(u, w^{\varepsilon}_{t+\Delta}, x^{\varepsilon}_{t+\Delta})]\\
&= \lim_{\Delta\to 0}\frac{1}{\Delta}\mathbb{E}[\varphi(u + C(u,x)\Delta + o(\Delta), w^{\varepsilon}_{t+\Delta}, x^{\varepsilon}_{t+\Delta}) - \varphi(u, w^{\varepsilon}_{t+\Delta}, x^{\varepsilon}_{t+\Delta})]\\
&= \lim_{\Delta\to 0}\frac{1}{\Delta}\mathbb{E}[\varphi(u, w^{\varepsilon}_{t+\Delta}, x^{\varepsilon}_{t+\Delta}) + \varphi'(u, w^{\varepsilon}_{t+\Delta}, x^{\varepsilon}_{t+\Delta})(C(u,x)\Delta + o(\Delta))\\
&\quad + o(\Delta) - \varphi(u, w^{\varepsilon}_{t+\Delta}, x^{\varepsilon}_{t+\Delta})]\\
&= \lim_{\Delta\to 0}\frac{1}{\Delta}\mathbb{E}[\varphi'(u, w^{\varepsilon}_{t+\Delta}, x^{\varepsilon}_{t+\Delta})(C(u,x)\Delta + o(\Delta)) + o(\Delta)]\\
&= C(u,x)\varphi'(u,w,x).
\end{aligned}$$

Similarly, from the relation for the generator $\Gamma^{\varepsilon}_{w}(x)$ and the famous relation for the generator of the Markov process

$$\lim_{\Delta\to 0}\frac{1}{\Delta}\mathbb{E}[\varphi(w, x^{\varepsilon}_{t+\Delta}) - \varphi(w,x)] = \varepsilon^{-1}\mathbf{Q}\varphi(w,x),$$

we obtain

$$\begin{aligned}
&\lim_{\Delta\to 0}\frac{1}{\Delta}\mathbb{E}[\varphi(u, w^{\varepsilon}_{t+\Delta}), x^{\varepsilon}_{t+\Delta}) - \varphi(u,w,x)]\\
&= \lim_{\Delta\to 0}\frac{1}{\Delta}\mathbb{E}[\varphi(u, w^{\varepsilon}_{t+\Delta}), x^{\varepsilon}_{t+\Delta})\\
&\quad - \varphi(u, w, x^{\varepsilon}_{t+\Delta}))] + \varphi(u, w, x^{\varepsilon}_{t+\Delta})) - \varphi(u,w,x)]\\
&= \lim_{\Delta\to 0}\frac{1}{\Delta}\mathbb{E}[\varphi(u, w^{\varepsilon}_{t+\Delta}), x^{\varepsilon}_{t+\Delta}) - \varphi(u, w, x^{\varepsilon}_{t+\Delta})]\\
&\quad + \lim_{\Delta\to 0}\frac{1}{\Delta}\mathbb{E}[\varphi(u, w, x^{\varepsilon}_{t+\Delta}) - \varphi(u,w,x)]\\
&= \Gamma^{\varepsilon}_{w}(x)\varphi(u,w,x) + \varepsilon^{-1}\mathbf{Q}\varphi(u,w,x),
\end{aligned}$$

and finally $\mathbf{L}^{\varepsilon}(x)$ has the form [2.23].

LEMMA 2.8.– *The generator [2.23] has the asymptotic representation*

$$\begin{aligned}
\mathbf{L}^{\varepsilon}(x)\varphi(u,w,x) &= \varepsilon^{-1}\mathbf{Q}\varphi(u,w,x)\\
&\quad + \Gamma^{1}_{u}(x)\varphi(u,w,x) + C(u)\varphi(u,w,x)\\
&\quad + \Gamma^{1}_{w}(x)\varphi(u,w,x) + \gamma^{\varepsilon}(x)\varphi(w,x)
\end{aligned}$$

where

$$C(u)\varphi(u) = C(u,x)\varphi'(u),$$

and the remainder term

$$\|\gamma^\varepsilon(x)\varphi(u,w,x)\| \to 0,\ \varepsilon \to 0,$$

$\varphi(u,w,\cdot) \in C^3((R))$.

The truncated generator has the form

$$\begin{aligned}\mathbf{L}_0^\varepsilon(x)\varphi(u,w,x) &= \varepsilon^{-1}\mathbf{Q}\varphi(u,w,x)\\ &+ \Gamma_u^1(x)\varphi(u,w,x) + C(u)\varphi(u,w,x)\\ &+ \Gamma_w^1(x)\varphi(u,w,x).\end{aligned} \quad [2.24]$$

LEMMA 2.9.– *The solution of the singular perturbation problem for the truncated generator [2.24] on the perturbed test function*

$$\varphi^\varepsilon(u,w,x) = \varphi(u,w) + \varepsilon\varphi_1(u,w,x) \quad [2.25]$$

is determined by the equality

$$L_0^\varepsilon(x)\varphi^\varepsilon(u,w,x) = \hat{L}\varphi(u,w) + \varepsilon\theta^\varepsilon(x)\varphi(u,w),$$

where

$$\begin{aligned}\theta^\varepsilon(x) &= \Gamma_u^1(x)R_0\hat{L} - \Gamma_u^1(x)R_0\Gamma_u^1(x)\\ &- \Gamma_u^1(x)R_0\mathbf{C}(x) - \Gamma_u^1(x)R_0\Gamma_w^1(x)\\ &+ \mathbf{C}(x)R_0\hat{L} - \mathbf{C}(x)R_0\Gamma_u^1(x)\\ &- \mathbf{C}(x)R_0\mathbf{C}(x) - \mathbf{C}(x)R_0\Gamma_w^1(x)\\ &+ \Gamma_w^1(x)R_0\hat{L} - \Gamma_w^1(x)R_0\Gamma_u^1(x)\\ &- \Gamma_w^1(x)R_0\mathbf{C}(x) - \Gamma_w^1(x)R_0\Gamma_w^1(x).\end{aligned}$$

PROOF.– By substituting [2.25] into [2.24], we have

$$\begin{aligned}\mathbf{L}_0^\varepsilon(x)\varphi(u,w,x) &= \varepsilon^{-1}\mathbf{Q}[\varphi(u,w) + \varepsilon\varphi_1(u,w,x)]\\ &+ \Gamma_u^1(x)[\varphi(u,w) + \varepsilon\varphi_1(u,w,x)]\\ &+ \mathbf{C}(x)[\varphi(u,w) + \varepsilon\varphi_1(u,w,x)]\\ &+ \Gamma_w^1[\varphi(u,w) + \varepsilon\varphi_1(u,w,x)]\end{aligned}$$

$$
\begin{aligned}
&= \varepsilon^{-1}\mathbf{Q}\varphi(u,w) + [\mathbf{Q}\varphi_1(u,w,x) + \Gamma_u^1(x)\varphi(u,w) \\
&\quad + \mathbf{C}(x)\varphi(u,v) + \Gamma_w^1(x)\varphi(u,w)] \\
&\quad + \varepsilon[\Gamma_u^1(x)\varphi_1(u,w,x) + \mathbf{C}(x)\varphi_1(u,w,x) \\
&\quad + \Gamma_w^1(x)\varphi_1(u,w,x)].
\end{aligned}
$$

For the existing limiting generator $\hat{L}\varphi(u,w)$, the relation

$$\mathbf{Q}\varphi(u,w) = 0,$$

holds true when $\varepsilon \to 0$, i.e. the function belongs to the zero-subspace of the operator $\mathbf{Q}$.

Then

$$\hat{L}\varphi(u,w) = \mathbf{Q}\varphi_1(u,w,x) + \Gamma_u^1(x)\varphi(u,w) + \mathbf{C}(x)\varphi(u,w) + \Gamma_w^1(x)\varphi(u,w),$$

where

$$\mathbf{Q}\varphi_1(u,w,x) = [\hat{L} - \Gamma_u^1(x) - \mathbf{C}(x) - \Gamma_w^1(x)]\varphi(u,w).$$

From the condition for resolvability of the last equation, we have

$$\Pi\mathbf{Q}\Pi\varphi_1(u,w,x) = 0 = \Pi[\hat{L} - \Gamma_u^1(x) - \mathbf{C}(x) - \Gamma_w^1(x)]\Pi\varphi(u,w).$$

Thus

$$
\begin{aligned}
&\hat{L}\varphi(u,w) = \Pi\Gamma_u^1(x)\varphi(u,w) + \Pi\mathbf{C}(x)\varphi(u,w) + \\
&+\Pi\Gamma_w^1(x)\varphi(u,w)
\end{aligned}
$$

and

$$\varphi_1(u,w,x) = R_0[\hat{L} - \Gamma_u^1(x) - \mathbf{C}(x) - \Gamma_w^1(x)]\varphi(u,w).$$

We obtain the following relation for the last equation:

$$
\begin{aligned}
&\varepsilon[\Gamma_u^1(x)\varphi_1(u,w,x) + \mathbf{C}(x)\varphi_1(u,w,x) + \Gamma_w^1(x)\varphi_1(u,w,x)] \\
&= \varepsilon[\Gamma_u^1(x)R_0[\hat{L} - \Gamma_u^1(x) - \mathbf{C}(x) - \Gamma_w^1(x)] \\
&\quad + \mathbf{C}[\hat{L} - \Gamma_u^1(x) - \mathbf{C}(x) - \Gamma_w^1(x)] \\
&\quad + \Gamma_w^1(x)R_0[\hat{L} - \Gamma_u^1(x) - \mathbf{C}(x) - \Gamma_w^1(x)]]\varphi(u,w).
\end{aligned}
$$

PROOF OF THEOREM 2.3.– According to conditions C1 – C6, the boundedness of the remainder term [2.26] holds true:

$$\begin{aligned}\|\theta^{\varepsilon}(x)V(u)\| = {} & |\Gamma_u^1(x)R_0\hat{L}V(u) - \Gamma_u^1(x)R_0\Gamma_u^1(x)V(u)\\ & - \Gamma_u^1(x)R_0\mathbf{C}(x)V(u) - \Gamma_u^1(x)R_0\Gamma_w^1(x)V(u)\\ & + \mathbf{C}(x)R_0\hat{L}V(u) - \mathbf{C}(x)R_0\Gamma_u^1(x)V(u)\\ & - \mathbf{C}(x)R_0\mathbf{C}(x)V(u) - \mathbf{C}(x)R_0\Gamma_w^1(x)V(u)\\ & + \Gamma_w^1(x)R_0\hat{L}V(u) - \Gamma_w^1(x)R_0\Gamma_u^1(x)V(u)\\ & - \Gamma_w^1(x)R_0\mathbf{C}(x)V(u) - \Gamma_w^1(x)R_0\Gamma_w^1(x)V(u)|\\ & \le M_1V(u) + M_2V(u) + M_3V(u) + M_4V(u)\\ & + M_5V(u) + M_6V(u),\end{aligned}$$

then

$$\|\theta^{\varepsilon}(x)V(u)\| \le MV(u), \tag{2.26}$$

where $M = M_1 + M_2 + M_3 + M_4 + M_5 + M_6$.

From Lemma 2.9, relation [2.26] and the Korolyuk model theorem (Korolyuk and Limnios 2005), we obtain the following weak convergence:

$$(u^{\varepsilon}(t), \eta^{\varepsilon}(t)) \Rightarrow (u(t), \eta(t)),\ \ \varepsilon \to 0.$$

Furthermore, let $\frac{d^{[2.1]}V(u)}{du}$ be a derivative of the Lyapunov function, which is calculated along the trajectory of the system [2.1]. Since the Lyapunov function must satisfy the Lipschitz condition

$$|V(u_2) - V(u_1)| < K|u_2 - u_1|,$$

where K is a constant, the following relation holds true:

$$\frac{d^{[2.1]}V(u)}{du} \le \frac{dV(u)}{du} + K[\|\sigma(u)\| \cdot |dW(t)| + \int_R v^2\tilde{\Gamma}_0(dv)|dt|],$$

where $\frac{d^{[2.1]}V(u)}{du}$ is a derivative of the Lyapunov function, which is calculated along the trajectory of the deterministic system [2.19]

$$\tilde{\Gamma}_0(dv) = \int_X \pi(dx)\Gamma_0(dv, x).$$

From conditions [2.20]–[2.22], we obtain

$$\frac{d^{[2.1]}V(u)}{du} \le -c_1V(u) + K[c_2|dW(t)| + C|dt|].$$

Thus, using Lemma 1.7 Nevelson and Khasminsky (1972), we have

$$V(u) \leq V(u_0)\exp\{-c_1 t\} + Kc_2 \int_0^t \exp\{-c_1(t-s)\}d|w(s)|ds + K\hat{c}_3|dt|.$$

From the last relation and Lemma 1.9 (Nevelson and Khasminsky 1972), we obtain the estimation

$$\mathbb{P}\{|u(t)| > R\} \leq \frac{V(u)}{\inf\limits_{u \in R^d} V(u)}, \quad R \to \infty,$$

Thus, system [2.1] is dissipative. Moreover, from the Korolyuk model theorem and the dissipativity of the limit process, we conclude that system [2.1] is asymptotic dissipative.

2.3. Double merging of phase space for differential equations with small stochastic supplements under Levy approximation conditions

We investigate the stochastic evolutionary system in an ergodic Markov environment

$$du^\varepsilon(t) = C(u^\varepsilon(t), x(t/\varepsilon^3))dt + d\eta^\varepsilon(t), \ u^\varepsilon(t) \in \mathbb{R}, \tag{2.27}$$

where $x^\varepsilon(t), t \geq 0$ is the Markov process determined on a standard phase space (E, E) with splitting

$$E = \bigcup_{k=1}^{N} E_k, \ E_k \cap E_{k'} = \emptyset, \ k \neq k'$$

in a series scheme with a small parameter of a series of $\varepsilon \to 0$, $\varepsilon > 0$.

The Markov kernel has the form

$$\mathbf{Q}^\varepsilon(x, B, t) = P^\varepsilon(x, B)[1 - \exp\{-q(x)t\}], \ x \in E, \ B \in E, \ t \geq 0.$$

Let us consider the following assumptions:

ME1: A kernel describing the transition probabilities of the embedded Markov chain $x_n^\varepsilon, n \geq 0$ is defined as follows:

$$P^\varepsilon(x, B) = P(x, B) + \varepsilon P_1(x, B).$$

The stochastic kernel $P(x, B)$ on the split phase space is defined as

$$P(x, E_k) = 1_k(x) = \begin{cases} 1, & x \in E_k, \\ 0, & x \notin E_k. \end{cases}$$

The stochastic kernel $P(x, B)$ defines the accompanying Markov chain $x_n, n \geq 0$ on classes E_k, $1 \leq k \leq N$. In addition, the perturbing kernel $P_1(x, B)$ satisfies the condition

$$P_1(x, E) = 0,$$

which is a direct corollary of the equality

$$P^\varepsilon(x, E) = P(x, E) = 1.$$

ME2: The associated Markov process $x^0(t)$, $t \geq 0$, given by the generator

$$\mathbf{Q}\varphi(x) = q(x) \int_E P(x, dy)[\varphi(y) - \varphi(x)],$$

is uniformly ergodic in each of the classes E_k, $1 \leq k \leq N$, with the stationary distribution $\pi_k(dx)$, $1 \leq k \leq N$, satisfying the relation

$$\pi_k(dx)q(x) = q_k \rho_k(dx), \ q_k := \int_{E_k} \pi_k(dx)q(x).$$

ME3: The average probability of exit

$$\hat{p}_k := q(x) \int_{E_k} \rho_k(dx)P_1(x, E/E_k) > 0, \ 1 \leq k \leq N.$$

Thus, the perturbing kernel $P_1(x, B)$ determines the transitional probabilities between classes E_k, $1 \leq k \leq N$; thus, the equality

$$P^\varepsilon(x, B) = P(x, B) + \varepsilon P_1(x, B)$$

means that the embedded Markov chain x_n^ε, $n \geq 0$ spends a great deal of time in each of the classes E_k and jumps between classes with low probabilities $\varepsilon P_1(x, E/E_k)$.

EXAMPLE 2.1.– *Consider a three-state Markov process, $E^0 = \{0, 1, 2\}$, with the generator matrix*

$$Q^\varepsilon = \begin{pmatrix} 0 & 0 & 0 \\ \varepsilon\lambda & -(1+\varepsilon)\lambda & \lambda \\ \varepsilon\mu & \mu & -(1+\varepsilon)\mu \end{pmatrix} = \begin{pmatrix} 0 & 0 & 0 \\ 0 & -\lambda & \lambda \\ 0 & \mu & -\mu \end{pmatrix} + \varepsilon \begin{pmatrix} 0 & 0 & 0 \\ \lambda & -\lambda & 0 \\ \mu & 0 & -\mu \end{pmatrix}.$$

The transition matrix of the embedded Markov chain is

$$P^\varepsilon = \begin{pmatrix} 1 & 0 & 0 \\ \varepsilon & 0 & 1-\varepsilon \\ \varepsilon & 1-\varepsilon & 0 \end{pmatrix} = \begin{pmatrix} 1 & 0 & 0 \\ 0 & 0 & 1 \\ 0 & 1 & 0 \end{pmatrix} + \varepsilon \begin{pmatrix} 0 & 0 & 0 \\ 1 & 0 & -1 \\ 1 & -1 & 0 \end{pmatrix}.$$

Now, for the ergodic process $x(t)$, $t \geq 0$, taking values in $E = \{1, 2\}$, and the generator Q, we have $\pi = (\frac{\mu}{\lambda+\mu}, \frac{\lambda}{\lambda+\mu})$. For the ergodic embedded Markov chain x_n, $n \geq 0$, we have $\rho = (1/2, 1/2)$.

Thus, since we have $p(1) = -P_1(1, E) = 1$ and $p(2) = -P_1(2, E) = 1$, the stoppage probability is $p = 1$.

However, we have $q(1) = \lambda$ and $q(2) = \mu$.

Hence

$$q = \pi_1 q(1) + \pi_2 q(2) = \frac{2\lambda}{\lambda + \mu}$$

and

$$\Lambda = qp = \frac{2\lambda\mu}{\lambda + \mu}.$$

The limit of the distribution of the normalized absorption time is

$$\mathbb{P}\{\zeta > t\} = \exp(-\Lambda t).$$

Under conditions ME1–ME3, there is a weak convergence (Korolyuk and Turbin 1993):

$$\nu(x^{\varepsilon}(t)) \Rightarrow \hat{x}(t), \ \varepsilon \to 0, \ \nu(x) = k \in \hat{E} = \{1, ..., N\}, \ x \in E_k, \ 1 \leq k \leq N.$$

The limited Markov process $\hat{x}(t), \ t \geq 0$ on the enlarged phase space $\hat{E} = \{1, ..., N\}$ is determined by the generator matrix

$$\hat{Q}_1 = (\hat{q}_{kr}, \ 1 \leq k, r \leq N),$$

where

$$\hat{q}_{kr} = \hat{q}_k \hat{p}_{kr}, \ k \neq r, \ \hat{q}_k = q_k \hat{p}_k, \ 1 \leq k \leq N,$$

$$\hat{p}_{kr} = p_{kr}/\hat{p}_k, \ p_{kr} = \int_{E_k} \rho_k(dx) P_1(x, E_r), \ 1 \leq k, r \leq N, \ k \neq r,$$

$$\hat{p}_k = -\int_{E_k} \rho_k(dx) P_1(x, E_k).$$

ME4: The merging Markov process $\hat{x}(t), \ t \geq 0$ is an ergodic process with the stationary distribution $\hat{\pi} = (\pi_k, \ k \in \hat{E})$.

Therefore, $\mathbf{Q}^{\varepsilon}$ is the operator that can be represented as

$$\mathbf{Q}^{\varepsilon} = \mathbf{Q} + \varepsilon \mathbf{Q}_1, \ \mathbf{Q}_1(x) = q(x) \int_E P_1(x, dy) \varphi(y).$$

The generalization of such approach can be found in Korolyuk and Korolyuk (1999), where $\mathbf{Q}^\varepsilon = \mathbf{Q} + \varepsilon\mathbf{Q}_1$:

$$\mathbf{Q}(x) = q(x)\int_E P(x,dy)[\varphi(y)-\varphi(x)],\ \mathbf{Q}_1(x) = q_1(x)\int_E P_1(x,dy)\varphi(y).$$

Let Π be the projector onto the zero-subspace of a reducible-invertible operator $\mathbf{Q}$. Its effect on the test functions is defined as follows:

$$\Pi\varphi(x) = \sum_{k=1}^{N} \hat{\varphi}_k 1_k(x),\ \hat{\varphi}_k := \int_{E_k} \pi_k(dx)\varphi(dx).$$

The reducible operator $\hat{Q}_1$ can be determined by the relation

$$\hat{Q}_1\Pi = \Pi\mathbf{Q}_1\Pi.$$

Let $\hat{\Pi}$ be the projector onto the zero-subspace of a reducible-invertible operator $\hat{Q}_1$:

$$\hat{\Pi}\hat{\varphi} := q(x)\sum_{k\in E} \hat{\pi}_k\hat{\varphi}_k.$$

The potential matrix $\hat{R}_0 = [\hat{R}^0_{kj};\ 1 \le k, l \le N]$ can be determined by the relation

$$\hat{Q}_1\hat{R}_0 = \hat{R}_0\hat{Q}_1 = \hat{\Pi} - E.$$

The IPP $\eta^\varepsilon(t), t \ge 0$, in the Levy approximation scheme can be determined by the relation

$$\eta^\varepsilon(t) = \int_0^t \eta^\varepsilon(ds, x(s/\varepsilon^3)), \tag{2.28}$$

where $\eta^\varepsilon(t,x)$, $t \ge 0$, $x \in X$ is a set of processes with independent increments that can be determined by generators

$$\Gamma^\epsilon(x)\varphi(w) = \varepsilon^{-2}\int_R (\varphi(w+v)-\varphi(w))\Gamma^\varepsilon(dv,x),\ x \in X \tag{2.29}$$

and satisfied with Levy approximation conditions.

L1. Approximation of averages

$$\int_R v\Gamma^\varepsilon(dv,x) = \varepsilon a_1(x) + \varepsilon^2(a_2(x)+\theta_a(x)),\ \theta_a(x)\to 0,\ \varepsilon\to 0,$$

and

$$\int_R v^2\Gamma^\varepsilon(dv,x) = \varepsilon(b(x)+\theta_b(x)),\ \theta_b(x)\to 0,\ \varepsilon\to 0,$$

L2. Condition for the distribution function

$$\int_R g(v)\Gamma^\varepsilon(dv,x) = \varepsilon^2(\Gamma_g(x)+\theta_g(x)),\ \theta_g(x)\to 0,\ \varepsilon\to 0,$$

for each $g(v)\in C^2(\mathbb{R})$ (space of bounded functions with values in $\mathbb{R}$ and $g(v)/|v|^2\to 0,\ |v|\to 0$). The measure $\Gamma_g(x)$ is bounded for each $g(v)\in C^2(\mathbb{R})$, which can be determined by the relation

$$\Gamma_g(x) = \int_R g(v)\Gamma_0(dv,x),\ \ g(v)\in C^2(\mathbb{R});$$

L3. Uniform quadratic integrability

$$\sup \lim_{c\to\infty} \int_{|v|>c} v^2\Gamma_0(dv,x) = 0;$$

Let us denote

$$\Gamma_1(x)\varphi(w) = a(x)\varphi'(w)$$

Let the condition of balance be fulfilled:

$$\hat{\Pi}\hat{\Gamma}_1 = 0, \qquad [2.30]$$

where $\hat{\Gamma}_1\hat{\varphi}(w) = \Pi\Gamma_1(x)\varphi(w)$.

We will further investigate the asymptotic properties of the perturbation process.

THEOREM 2.4.– *Let the balance condition [2.30] and L1 – L3 hold. Then, the weak convergence*

$$\eta^\varepsilon(t)\to\eta^0(t),\ \varepsilon\to 0$$

holds true.

The limit process $\eta^0(t)$ is determined by the generator

$$\hat{\hat{\Gamma}}\varphi(w) = \hat{a}_2\varphi'(w) + \frac{1}{2}\hat{\sigma}^2\varphi''(w) + \int_R [\varphi(w+v)-\varphi(w)]\hat{\hat{\Gamma}}_0(dv),$$

where

$$\hat{a}_2 = \sum_{k \in \hat{E}} \hat{\pi}_k \int_{E_k} \pi_k(dx)(a_2(x) - a_0(x)),$$

$$\hat{\sigma}^2 = \sum_{k \in \hat{E}} \hat{\pi}_k \int_{E_k} \pi(dx)(b(x) - b_0(x)) + 2 \sum_{k \in \hat{E}} \hat{\pi}_k \int_{E_k} \pi(dx) a_1(x) R_0 a_1(x),$$

$$\hat{a}_0(x) = \sum_{k \in \hat{E}} \hat{\pi}_k \int_{E_k} v\Gamma_0(dv, x),$$

$$\hat{b}_0(x) = \sum_{k \in \hat{E}} \hat{\pi}_k \int_{E_k} v^2 \Gamma_0(dv, x),$$

$$\hat{\Gamma}_0(v) = \sum_{k \in \hat{E}} \hat{\pi}_k \int_{E_k} \pi(dx)\Gamma_0(v, x)$$

which is a Levy process that has three components: deterministic shift, diffusion component and Poisson jump component.

PROOF.– First, let us prove some additional propositions.

LEMMA 2.10.– *Generators of independent increment processes $\eta^\varepsilon(t, x)$, $t \geq 0$, $x \in X$, acting on test functions $\varphi(w) \in C^2(\mathbf{R})$ under assumptions L1 – L3 can have the asymptotic presentation*

$$\Gamma^\varepsilon(x)\varphi(w) = \varepsilon^{-1}\Gamma_1(x)\varphi(w) + \Gamma_2(x)\varphi(w), \qquad [2.31]$$

where

$$\Gamma_1(x)\varphi(w) = a_1(x)\varphi'(w),$$

$$\Gamma_2(x)\varphi(w) = (a_2(x) - a_0(x))\varphi'(x) + \frac{1}{2}(b(x) - b_0(x))\varphi''(x) + \int_R [\varphi(w + v) - \varphi(v)]\Gamma_0(dv, x).$$

PROOF OF LEMMA 2.10.– We use the expansion of the function $\varphi(w)$ in the Taylor series. From [2.29], we then obtain

$$\mathbf{\Gamma}^\varepsilon(x)\varphi(w) = \varepsilon^{-2} \int_R (\varphi(w + v) - \varphi(v))\Gamma^\varepsilon(dv, x)$$

$$= \varepsilon^{-2} \int_R (\varphi(w+v) - \varphi(v) - v\varphi'(v) - \frac{1}{2}v^2\varphi''(w))\Gamma^\varepsilon(dv, x)$$

$$+\varepsilon^{-2} \int_R (v\varphi'(w)\Gamma^\varepsilon(dv, x) + \frac{1}{2}v^2\varepsilon^{-2} \int_R v^2\varphi''(w)\Gamma^\varepsilon(dv, x)$$

$$= \int_R (\varphi(u+v) - \varphi(v) - v\varphi'(w) - \frac{1}{2}v^2\varphi''(w))\Gamma_0(dv, x)$$

$$+\varepsilon^{-1}a_1(x)\varphi'(w) + a_2(x)\varphi'(w) + \frac{1}{2}(b(x) - b_0(x))\varphi''(w)$$

$$+ \int_R (\varphi(u+v) - \varphi(v))\Gamma_0(dv, x) + \gamma^\varepsilon(w)\varphi(w),$$

where the penultimate equality follows from L1–L3. We also note that the function $\varphi(w+v) - \varphi(w) - v\varphi'(w) - \frac{1}{2}v^2\varphi''(w) \in C^2(\mathbb{R})$, because it is bounded and $\varphi(w)$ is bounded along with its derivatives, and

$$[\varphi(w+v) - \varphi(w) - v\varphi'(w) - \frac{1}{2}v^2\varphi''(w)]/|v^2| \to 0$$

when $v \to 0$.

We recall that $\gamma^\varepsilon(w)\varphi(w) = o(\varepsilon^2)$, $\varphi(w) \in C^2(\mathbb{R})$; thus, we obtain the representation [2.31].

LEMMA 2.11.– *The generator of a two-component Markov process* $(\eta^\varepsilon, x(t/\varepsilon^2))$, $t \geq 0$, *can be represented as follows:*

$$\hat{\Gamma}^\varepsilon(x)\varphi(w, x) = \varepsilon^{-3}\mathbf{Q}\varphi(w, x) + \varepsilon^{-1}\Gamma_1(x)\varphi(w, x)$$
$$+\Gamma_2(x)\varphi(w, x) + \gamma^\varepsilon(x)\varphi(w, x), \quad [2.32]$$

where $\Gamma_1(x)$ *and* $\Gamma_2(x)$ *are defined in Lemma 2.10 and the remainder term* $\|\gamma^\varepsilon(x)\varphi(w, x)\| \to 0$ *when* $\varepsilon \to 0$, $\varphi(w, \cdot) \in C^2(\mathbb{R})$.

PROOF.– The assertion of the lemma follows from the definition of the Markov process generator as well as from the form of the corresponding process generators $\eta^\varepsilon(t, x)$ and $x(t/\varepsilon^2)$.

The cut operator has the form

$$\mathbf{L}^\varepsilon = \varepsilon^{-3}\mathbf{Q}^\varepsilon + \varepsilon^{-1}\Gamma_1 + \Gamma_2 = \varepsilon^{-3}\mathbf{Q} + \varepsilon^{-2}\mathbf{Q}_1 + \varepsilon^{-1}\Gamma_1 + \Gamma_2. \quad [2.33]$$

LEMMA 2.12.– *Under the balance condition [2.30], the solution of the singular perturbation problem for the cut operator [2.33] on test functions*

$$\varphi(u,x) = \varphi(u) + \varepsilon\varphi_1 + \varepsilon^2\varphi_2 + \varepsilon^3\varphi_3$$

is determined by the relation

$$\Gamma_0^\varepsilon(x)\varphi^\varepsilon(u,x) = \hat{\hat{L}}\varphi(u) + \varepsilon\theta_\eta^\varepsilon(x)\varphi(u), \qquad [2.34]$$

where the remainder term is uniformly bounded with respect to x.

The limiting operator can be represented by the formula

$$\hat{\hat{L}} = \hat{\Pi}\hat{\Gamma}_1\hat{R}_0\hat{\Gamma}_1\hat{\Pi} + \hat{\Pi}\hat{\Gamma}_2\hat{\Pi}. \qquad [2.35]$$

PROOF.– Let us calculate

$$(\varepsilon^{-3}\mathbf{Q} + \varepsilon^{-2}\mathbf{Q}_1 + \varepsilon^{-1}\Gamma_1 + \Gamma_2)(\varphi + \varepsilon\varphi_1 + \varepsilon^2\varphi_2 + \varepsilon^3\varphi_3)$$
$$= \varepsilon^{-3}\mathbf{Q}\varphi + \varepsilon^{-2}(\mathbf{Q}\varphi_1 + \mathbf{Q}_1\varphi) + \varepsilon^{-1}(\mathbf{Q}\varphi_2 + \mathbf{Q}_1\varphi_1 + \Gamma_1\varphi)$$
$$+(\mathbf{Q}\varphi_3 + \mathbf{Q}_1\varphi_2 + \Gamma_1\varphi_1 + \Gamma_2\varphi) + o(\varepsilon).$$

From the above, we obtain four relations:

$$\mathbf{Q}\varphi = 0; \qquad [2.36]$$

$$\mathbf{Q}\varphi_1 + \mathbf{Q}_1\varphi = 0; \qquad [2.37]$$

$$\mathbf{Q}\varphi_2 + \mathbf{Q}_1\varphi_1 + \Gamma_1\varphi = 0; \qquad [2.38]$$

$$\mathbf{Q}\varphi_3 + \mathbf{Q}_1\varphi_2 + \Gamma_1\varphi_1 + \Gamma_2\varphi = \hat{\hat{L}}\varphi. \qquad [2.39]$$

Furthermore, we define the form of $\hat{\hat{L}}$.

From [2.36], it follows that $\varphi \in N_Q$.

According to [2.37], since $\varphi \in N_Q$, from the resolvability condition, we have

$$\Pi Q_1 \Pi\varphi = 0.$$

Let us denote

$$\Pi\mathbf{Q}_1 = \hat{Q}_1, \ \Pi\varphi = \hat{\varphi}.$$

Then

$$\hat{Q}_1\hat{\varphi} = 0,$$

furthermore

$$\hat{\varphi} \in N_{\hat{Q}_1}.$$

Let us consider [2.38]: from the resolvability condition for $\mathbf{Q}$, we obtain

$$\Pi \mathbf{Q}_1 \Pi \varphi_1 + \Pi \Gamma_1 \Pi \varphi = 0, \qquad [2.40]$$

$$\hat{Q}_1 \hat{\varphi}_1 + \hat{\Gamma}_1 \hat{\varphi} = 0.$$

According to the balance condition [2.30], we have $\hat{\Gamma}_1 \hat{\varphi} \in R_Q$, so the solution

$$\hat{\varphi}_1 = \hat{R}_0 \hat{\Gamma}_1 \hat{\varphi},$$

where $\hat{R}_0$ is reducible-invertible to $\hat{Q}_1$.

Let us consider [2.39]: from the resolvability condition for $\mathbf{Q}$, we obtain

$$\Pi \mathbf{Q} \Pi \varphi_2 + \Pi \Gamma_1 \Pi \varphi_1 + \Pi \Gamma_2 \Pi \varphi = \Pi \hat{\hat{L}} \Pi \varphi, \qquad [2.41]$$

$$\hat{Q}_1 \hat{\varphi}_2 + \hat{\Gamma}_1 \hat{\varphi}_1 + \hat{\Gamma}_2 \hat{\varphi} = \hat{\hat{L}} \hat{\varphi}.$$

Recalling $\hat{\varphi}_1 = \hat{R}_0 \hat{\Gamma}_1 \hat{\varphi}$, we have

$$\hat{Q}_1 \hat{\varphi}_2 + \hat{\Gamma}_1 \hat{R}_0 \hat{\Gamma}_1 \hat{\varphi} + \hat{\Gamma}_2 \hat{\varphi} = \hat{\hat{L}} \hat{\varphi}.$$

Furthermore, from the resolvability condition for $\hat{\varphi}_2$

$$\hat{\Pi} \hat{\Gamma}_1 \hat{R}_0 \hat{\Gamma}_1 \hat{\Pi} \varphi + \hat{\Pi} \hat{\Gamma}_2 \hat{\Pi} \hat{\varphi} = \hat{\hat{L}} \hat{\varphi},$$

where

$$\hat{\hat{L}} = \hat{\Pi} \hat{\Gamma}_1 \hat{R}_0 \hat{\Gamma}_1 \hat{\Pi} + \hat{\Pi} \hat{\Gamma}_2 \hat{\Pi},$$

$$\hat{\varphi}_2 = \hat{R}_0 [\hat{\Gamma}_1 \hat{R}_0 \hat{\Gamma}_1 + \hat{\Gamma}_2 - \hat{\hat{L}}] \hat{\varphi},$$

$$\hat{\varphi}_3 = R_0 [Q_1 \varphi_2 + \Gamma_1 \varphi_1 + \Gamma_2 \varphi - \hat{\hat{L}} \varphi].$$

The boundedness of $\theta_\eta^\varepsilon(x)\varphi(w)$ follows from the form of operators Γ_1, Γ_2 and R_0.

The proof of the theorem is completed using Lemma 3 and Theorem 6.3 in Korolyuk and Limnios (2005).

Furthermore, let us consider the asymptotic properties of the input evolutionary system [2.27].

THEOREM 2.5.– *Under the balance condition [2.30], the weak convergence*

$$(u^\varepsilon(t), \eta^\varepsilon(t)) \Rightarrow (\hat{u}(t), \eta^0(t)), \ \varepsilon \to 0$$

holds true.

The limiting process $(\hat{u}(t), \eta^0(t))$ *is given by the generator*

$$\mathbf{L}\varphi(u,w) = \hat{\hat{C}}(u)\varphi'_u(u,w) + \hat{\hat{\Gamma}}_1^u\varphi(u,\cdot) + \hat{\hat{\Gamma}}_1^w\varphi(\cdot,w)\hat{R}_0\hat{\hat{\Gamma}}_1^w\varphi(\cdot,w) + \hat{\hat{\Gamma}}_2^w\varphi(\cdot,w) \quad [2.42]$$

where

$$\hat{\hat{C}}(u) = \Pi\mathbf{C}(x) = \int_X \pi(dx)C(u,x);$$

and the generators $\hat{\hat{\Gamma}}_1^u$ *and* $\hat{\hat{\Gamma}}_{1,2}^w$ *are determined using Lemma 2.10, which have the same form, but they act at different variables.*

NOTE 2.4.– *The weak convergence of processes* $(u^\varepsilon(t), \eta^\varepsilon(t)) \Rightarrow (\hat{u}(t), \eta^0(t))$, $\varepsilon \to 0$, *follows from the convergence of the respective generators when the prelimiting set of processes* $u^\varepsilon(t)$ *is compact. Theorems on the compactness of processes with independent increments in the Levy approximation scheme were proved, in particular, in [4].*

PROOF OF THEOREM 2.5.–

LEMMA 2.13.– *The generator of a three-component Markov process* $(u^\varepsilon(t), \eta^\varepsilon(t), x^\varepsilon(t/\varepsilon^3))$, $t \geq 0$, *can be represented as follows:*

$$\begin{aligned}\mathbf{L}^\varepsilon(x)\varphi(u,w,x) &= \varepsilon^{-3}\mathbf{Q}^\varepsilon\varphi(u,w,x) + \Gamma_u^\varepsilon(x)\varphi(u,\cdot,x) + \varepsilon\Gamma_w^\varepsilon(x)\varphi(\cdot,w,x) \\ &+ \mathbf{C}(x)\varphi(u,w,x) + \theta_w^\varepsilon(x)\varphi(u,w,x),\end{aligned} \quad [2.43]$$

where $\Gamma^\varepsilon(x)$ *is the generator of the set of IPP [2.29],*

$$\mathbf{C}(x)\varphi(u,w,x) = C(u,x)\varphi'_u(u,w,x).$$

The remainder term $\|\theta_w^\varepsilon(x)\varphi(u,w,x)\| \to 0$ *when* $\varepsilon \to 0$.

The proof of the lemma can be found in Anisimov (2008).

LEMMA 2.14.– *The generator* $\mathbf{L}^\varepsilon(x)$ *in the case of the IPP has the asymptotic representation*

$$\begin{aligned}\mathbf{L}^\varepsilon(x)\varphi(u,w,x) &= \varepsilon^{-3}\mathbf{Q}^\varepsilon\varphi(u,w,x) + \varepsilon^{-1}\Gamma_1^w(x)\varphi(u,w,x) \\ &+ \Gamma_2^w(x)\varphi(u,w,x) + \Gamma_1^u(x)\varphi(u,w,x) \\ &+ \mathbf{C}(x)\varphi(u,w,x) + \hat{\theta}_w^\varepsilon\varphi(u,w,x),\end{aligned} \quad [2.44]$$

where

$$\hat{\theta}^\varepsilon_w(x) = \gamma^\varepsilon + \theta^\varepsilon_w(x),$$

$\Gamma_1^{\cdot}(x)$ *and* $\Gamma_2^{\cdot}(x)$ *are given in Lemma 2.10.*

The remainder term $\|\hat{\theta}^\varepsilon_w(x)\varphi(u,w,x)\| \to 0$ *when* $\varepsilon \to 0$.

PROOF.– The proof is based on the representation of the operator [2.31] and the results of Lemma 2.13.

The truncated operator has the form:

$$\mathbf{L}^\varepsilon(x)\varphi(u,w,x) = \varepsilon^{-3}\mathbf{Q}^\varepsilon\varphi(u,w,x) + \varepsilon^{-1}\Gamma_1^w(x)\varphi(u,w,x) + \Gamma_2^w(x)\varphi(u,w,x) + \Gamma_1^u(x)\varphi(u,w,x) + \mathbf{C}(x)\varphi(u,w,x) \quad [2.45]$$

LEMMA 2.15.– *Under the balance condition [2.30], the solution of the singular perturbation problem for the cut operator [2.45] on test functions*

$$\varphi^\varepsilon(w,x) = \varphi(w) + \varepsilon\varphi_1(w,x) + \varepsilon^2\varphi_2(w,x) + \varepsilon^3\varphi_3(w,x)$$

can be determined from the relation

$$\mathbf{L}_0^\varepsilon(x)\varphi^\varepsilon(w,x) = \mathbf{L}\varphi(w) + \varepsilon^3\theta^\varepsilon_w(x)\varphi(w), \quad [2.46]$$

where the remainder term $\theta^\varepsilon_w(x)$ *is uniformly bounded with respect to* x.

The limiting operator $\mathbf{L}$ *is given by the formula*

$$\mathbf{L} = \Pi[\hat{\hat{C}} + \hat{\hat{\Gamma}}_1^u + \hat{\hat{\Gamma}}_1^w \hat{R}_0 \hat{\Gamma}_1^w + \hat{\hat{\Gamma}}_2^w]\Pi. \quad [2.47]$$

PROOF.– To prove the equality [2.46], it is necessary that the coefficients at the same degrees ε on the left and the right be equal. Therefore, we can calculate

$$(\varepsilon^{-3}\mathbf{Q} + \varepsilon^{-2}\mathbf{Q}_2 + \varepsilon^{-1}\Gamma_1^u + \Gamma_2^u + \varepsilon^{-1}\Gamma_1^w + \Gamma_2^w + \mathbf{C})(\varphi + \varepsilon\varphi_1 + \varepsilon^2\varphi_2 + \varepsilon^3\varphi_3)$$
$$= \varepsilon^{-3}\mathbf{Q}\varphi + \varepsilon^{-2}(\mathbf{Q}\varphi_1 + \mathbf{Q}_1\varphi) + \varepsilon^{-1}(\mathbf{Q}\varphi_2 + \mathbf{Q}_1\varphi_1 + \Gamma_1^u\varphi + \Gamma_1^w\varphi)$$
$$+ (\mathbf{Q}\varphi_3 + \mathbf{Q}_1\varphi_2 + \Gamma_1^u\varphi_1 + \Gamma_1^w\varphi_1 + \Gamma_2^u\varphi + \Gamma_2^w\varphi + \mathbf{C}\varphi) + o(\varepsilon).$$

Again, we obtain four relations:

$$\mathbf{Q}\varphi = 0; \quad [2.48]$$

$$\mathbf{Q}\varphi_1 + \mathbf{Q}_1\varphi = 0; \quad [2.49]$$

$$\mathbf{Q}\varphi_2 + \mathbf{Q}_1\varphi_1 + \Gamma_1^u\varphi + \Gamma_1^w\varphi = 0; \qquad [2.50]$$

$$\mathbf{Q}\varphi_3 + \mathbf{Q}_1\varphi_2 + \Gamma_1^u\varphi_1 + \Gamma_1^w\varphi_1 + \Gamma_2^u\varphi + \Gamma_2^w\varphi + \mathbf{C}\varphi = \hat{\hat{L}}\varphi. \qquad [2.51]$$

Let us define the form of $\hat{\hat{L}}$.

From [2.48], it follows that $\varphi \in N_Q$;

According to [2.49], since $\varphi \in N_Q$, from the resolvability condition, we have

$$\Pi\mathbf{Q}_1\Pi\varphi = 0.$$

Let us denote

$$\Pi\mathbf{Q}_1 = \hat{Q}_1,\ \Pi\varphi = \hat{\varphi}.$$

then

$$\hat{Q}_1\hat{\varphi} = 0,$$

and

$$\hat{\varphi} \in N_{\hat{Q}_1}.$$

According to [2.50], from the resolvability condition for $\mathbf{Q}$, we have

$$\Pi\mathbf{Q}_1\Pi\varphi_1 + \Pi\Gamma_1\Pi\varphi = 0,$$

$$\hat{Q}_1\hat{\varphi}_1 + \hat{\Gamma}_1^u\varphi + \hat{\Gamma}_1^w\hat{\varphi} = 0.$$

From the balance condition [2.30], we have $\hat{\Gamma}_1^u\hat{\varphi}, \hat{\Gamma}_1^u\hat{\varphi} \in N_Q$, so the solution

$$\hat{\varphi}_1 = \hat{R}_0[\hat{\Gamma}_1^u + \hat{\Gamma}_1^w]\hat{\varphi},$$

where $\hat{R}_0$ is reducible-invertible to $\hat{Q}_1$.

According to [2.51], from the resolvability condition for $\mathbf{Q}$, we have

$$\Pi Q_2\Pi\varphi_2 + \Pi\Gamma_1^u\Pi\varphi_1 + \Pi\Gamma_2^u\Pi\varphi + \Pi\Gamma_1^w\Pi\varphi_1 + \Pi\Gamma_2^w\Pi\varphi + \Pi\mathbf{C}\Pi\varphi = \Pi\hat{\hat{L}}\Pi\varphi,$$

$$\hat{Q}_1\hat{\varphi}_2 + \hat{\Gamma}_1^u\hat{\varphi}_1 + \hat{\Gamma}_1^w\hat{\varphi}_1 + \hat{\Gamma}_2^u\hat{\varphi} + \hat{\Gamma}_2^w\hat{\varphi} + \hat{C}\hat{\varphi} = \hat{\hat{L}}\hat{\varphi}.$$

Recalling $\hat{\varphi}_1 = \hat{R}_0\hat{\Gamma}_1^u\hat{\varphi} + \hat{R}_0\hat{\Gamma}_1^w\hat{\varphi}$, we obtain

$$\hat{Q}_1\hat{\varphi}_2 + \hat{\Gamma}_1^u\hat{R}_0\hat{\Gamma}_1^u\hat{\varphi} + \hat{\Gamma}_1^u\hat{R}_0\hat{\Gamma}_1^w\hat{\varphi}$$

$$+\hat{\Gamma}_1^w\hat{R}_0\hat{\Gamma}_1^u\hat{\varphi} + \hat{\Gamma}_1^w\hat{R}_0\hat{\Gamma}_1^w\hat{\varphi} + \hat{\Gamma}_2^u\hat{\varphi} + \hat{\Gamma}_2^w\hat{\varphi} + \hat{C}\hat{\varphi} = \hat{\hat{L}}\hat{\varphi}.$$

Furthermore, from the resolvability condition for $\hat{\varphi}_2$

$$\hat{\Pi}\hat{\Gamma}_1^u \hat{R}_0 \hat{\Gamma}_1^u \Pi\hat{\varphi} + \hat{\Pi}\hat{\Gamma}_1^u \hat{R}_0 \hat{\Gamma}_1^w \Pi\hat{\varphi} + \hat{\Pi}\hat{\Gamma}_1^w \hat{R}_0 \hat{\Gamma}_1^u \Pi\hat{\varphi}$$

$$+\hat{\Pi}\hat{\Gamma}_1^w \hat{R}_0 \hat{\Gamma}_1^w \Pi\hat{\varphi} + \hat{\Pi}\hat{\Gamma}_2^u \Pi\hat{\varphi} + \hat{\Pi}\hat{\Gamma}_2^w \Pi\hat{\varphi} + \hat{\Pi}\hat{C}\hat{\Pi}\hat{\varphi} = \hat{\hat{L}}\hat{\hat{\varphi}}.$$

where

$$\hat{\hat{L}} = \hat{\Pi}\hat{\Gamma}_1 \hat{R}_0 \hat{\Gamma}_1 \hat{\Pi} + \hat{\Pi}\hat{\Gamma}_2 \hat{\Pi} + \hat{\Pi}\hat{C}\hat{\Pi},$$

$$\hat{\varphi}_2 = \hat{R}_0[\hat{\hat{L}} - \hat{\Gamma}_1 \hat{R}_0 \hat{\Gamma}_1 - \hat{\Gamma}_2 - \hat{C}]\hat{\varphi},$$

$$\hat{\varphi}_3 = \hat{R}_0[\hat{\hat{L}} + \mathbf{Q}_1\varphi_2 + \Gamma_1\varphi_1 + \hat{\Gamma}_2\varphi + \mathbf{C}\varphi].$$

The boundedness of $\theta_\eta^\varepsilon(x)\varphi(w)$ follows from the form of the operators $\hat{\Gamma}_1$, $\hat{\Gamma}_2$ and R_0.

The proof of the theorem is completed using Lemma 3 and Theorem 6.3 in Korolyuk and Limnios (2005).

3

Asymptotical Analysis of Random Evolutionary Systems Under Poisson Approximation Conditions

3.1. Differential equations with small stochastic additions under Poisson approximation conditions

In this section, we will consider the case where system disturbances are defined by an impulse process in the Poisson nonclassical approximation scheme (Korolyuk and Limnios 2005)

The stochastic evolutionary system (Samoilenko *et al.* 2017) in an ergodic Markov environment is defined by the stochastic differential equation

$$du^{\varepsilon}(t) = C(u^{\varepsilon}, x(t/\varepsilon))dt + d\eta^{\varepsilon}(t), \ u^{\varepsilon}(t) \in \mathbf{R}, \qquad [3.1]$$

where a uniformly ergodic Markov process $x(t)$ in a standard phase space $(X, \mathbf{X})$ is defined by the generator

$$\mathbf{Q}\varphi(x) = q(x) \int_X P(x, dy)[\varphi(y) - \varphi(x)]$$

on the Banach space $B(X)$ of real-valued bounded functions $\varphi(x)$ with the supremum norm

$$\|\varphi\| = \max_{x \in X} |\varphi(x)|.$$

The stochastic kernel $P(x, B)$, $x \in X$, $B \in \mathbf{X}$ defines uniformly ergodic embedded Markov chain $x_n = x(\tau_n)$, $n \geq 0$, with stationary distribution $\rho(B)$,

$B \in \mathbf{X}$. The stationary distribution $\pi(B)$, $B \in \mathbf{X}$ of the Markov process $x(t)$, $t \geq 0$, is defined by the relation

$$\pi(dx)q(x) = q\rho(dx), \quad q = \int_X \pi(dx)q(x).$$

We denote by R_0 the potential operator of the generator $\mathbf{Q}$, which is defined by the equality (Korolyuk and Limnios 2005)

$$R_0 = \Pi - (\Pi + \mathbf{Q})^{-1},$$

where $\Pi\varphi(x) = \int_X \pi(dy)\varphi(y)1(x)$ is the projector of zeroes of the operator $\mathbf{Q}$ onto the subspace $N_Q = \{\varphi : \ \mathbf{Q}\varphi = 0\}$.

The impulse perturbation process (IPP) $\eta^\varepsilon(t)$, $t \geq 0$, in the Poisson approximation scheme is defined by the relation

$$\eta^\varepsilon(t) = \int_0^t \eta^\varepsilon(ds, x(s/\varepsilon)), \qquad [3.2]$$

where the family of processes with independent increments $\eta^\varepsilon(t, x)$, $t \geq 0$, $x \in X$, is defined by the generators

$$\Gamma^\epsilon(x)\varphi(w) = \varepsilon^{-1} \int_R (\varphi(w+v) - \varphi(w))\Gamma^\varepsilon(dv, x), \quad x \in X \qquad [3.3]$$

and satisfies the properties of the Poisson approximation (Korolyuk and Limnios 2005).

P1 is the approximation in averages

$$\int_R v\Gamma^\varepsilon(dv, x) = \varepsilon(a(x) + \theta_a(x)), \ \theta_a(x) \to 0, \ \varepsilon \to 0,$$

$$\int_R v^2\Gamma^\varepsilon(dv, x) = \varepsilon(b(x) + \theta_b(x)), \ \theta_b(x) \to 0, \ \varepsilon \to 0;$$

P2 is the condition imposed on the distributing function

$$\int_R g(v)\Gamma^\varepsilon(dv, x) = \varepsilon(\Gamma_g(x) + \theta_g(x)), \ \theta_g(x) \to 0, \ \varepsilon \to 0,$$

for all $g(v) \in C^2(\mathbb{R})$ (the space of real-valued bounded functions such that $g(v)/|v|^2 \to 0$, $|v| \to 0$), where the measure $\Gamma_g(x)$ is bounded for all $g(v) \in C^2(\mathbb{R})$ (functions from the space $C^2(\mathbb{R})$ separate the measures):

$$\Gamma_g(x) = \int_R g(v)\Gamma_0(dv, x), \;\; g(v) \in C^2(\mathbb{R});$$

P3 is the uniform quadratic integrability of

$$\sup \lim_{c \to \infty} \int_{|v|>c} v^2\Gamma_0(dv, x) = 0;$$

P4 is the absence of the diffusion component

$$b(x) = \int_R v^2\Gamma_0(dv, x).$$

Let us introduce the notation

$$\Gamma_1(x) = a(x)\varphi'(w) + \int_R [\varphi(w+v) - \varphi(v) - v\varphi'(w)]\Gamma_0(dv, x).$$

Consider the asymptotic properties of the perturbed process.

THEOREM 3.1.– *If conditions P1 – P4 are satisfied, the weak convergence*

$$\eta^\varepsilon(t) \to \eta^0(t), \;\; \varepsilon \to 0$$

takes place for IPP [3.2]. The limiting process $\eta^0(t)$ is defined by the generator

$$\begin{aligned}\Gamma\varphi(w) &= \Pi\Gamma_1(x)\varphi(w) \\ &= \tilde{a}\varphi'(w) + \int_R [\varphi(w+v) - v\varphi'(w)]\tilde{\Gamma}_0(dv),\end{aligned}$$

where $\tilde{a} = \int_X \pi(dx)a(x)$, $\tilde{\Gamma}_0(v) = \int_X \pi(dx)\Gamma_0(v, x)$, and is a process with independent increments, which has both Poisson component and deterministic shift.

PROOF.– Before directly proving Theorem 3.1, we consider some propositions.

LEMMA 3.1.– *Generators of processes with independent increments $\eta^\varepsilon(t, x)$, $t \geq 0$, $x \in X$, on test functions $\varphi(w) \in C^2(\mathbf{R})$ under conditions P1 – P4 have the asymptotic representation*

$$\mathbf{\Gamma}^\varepsilon(x)\varphi(w) = \Gamma_1(x)\varphi(w) + \gamma^\varepsilon(x)\varphi(w), \qquad [3.4]$$

where

$$\Gamma_1(x)\varphi(w) = a(x)\varphi'(w) + \int_R [\varphi(w+v) - \varphi(v) - v\varphi'(w)]\Gamma_0(dv, x),$$

and the remainder term $\|\gamma^\varepsilon(x)\varphi(w)\| \to 0$ *as* $\varepsilon \to 0$, $\varphi(w,\cdot) \in C^2(\mathbf{R})$.

PROOF.– Using the expansion of the function $\Gamma_1(x)\varphi(w)$ into the Taylor series, let us transform the generator [3.3]:

$$\begin{aligned}
\Gamma^\varepsilon(x)\varphi(w) &= \varepsilon^{-1}\int_R (\varphi(w+v) - \varphi(v))\Gamma^\varepsilon(dv, x) \\
&= \varepsilon^{-1}\int_R (\varphi(w+v) - \varphi(v) - v\varphi'(v) - \frac{1}{2}v^2\varphi''(w))\Gamma^\varepsilon(dv, x) \\
&\quad + \varepsilon^{-1}\int_R (v\varphi'(w)\Gamma^\varepsilon(dv, x) + \frac{1}{2}v^2\varepsilon^{-1}\int_R v^2\varphi''(w)\Gamma^\varepsilon(dv, x) \\
&= \int_R (\varphi(u+v) - \varphi(v) - v\varphi'(w) - \frac{1}{2}v^2\varphi''(w))\Gamma_0(dv, x) \\
&\quad + a(x)\varphi'(w) + \frac{1}{2}b(x)\varphi''(w) + \gamma^\varepsilon(x)\varphi(w) \\
&= \int_R (\varphi(u+v) - \varphi(v) - v\varphi'(w))\Gamma_0(dv, x) \\
&\quad + a(x)\varphi'(w) + \gamma^\varepsilon(w)\varphi(w),
\end{aligned}$$

where the penultimate equality follows from conditions P1 and P2. Note that the function $\varphi(w+v) - \varphi(w) - v\varphi'(w) - \frac{1}{2}v^2\varphi''(w) \in C^3(\mathbf{R})$, since it is bounded due to the boundedness of $\varphi(w)$ and of its derivatives and $[\varphi(w+v) - \varphi(w) - v\varphi'(w) - \frac{1}{2}v^2\varphi''(w)]/|v^2| \to 0,\ v \to 0)$, and the last equality follows from condition P4.

Taking into account that $\gamma^\varepsilon(w)\varphi(w) = O(\varepsilon^2)$, $\varphi(w) \in C^2(\mathbf{R})$, we obtain representation [3.4].

Lemma 3.1 is proved.

LEMMA 3.2.– *The generator of a two-component Markov process* $(\eta^\varepsilon, x(t/\varepsilon))$ *has the form*

$$\hat{\Gamma}^\varepsilon(x)\varphi(w,x) = \varepsilon^{-1}\mathbf{Q}\varphi(w,x) + \Gamma_1(x)\varphi(w,x) + \gamma^\varepsilon(x)\varphi(w,x), \qquad [3.5]$$

where the operator $\Gamma_1(x)$ *is defined in Lemma 3.1, and the remainder term* $\|\gamma^\varepsilon(x)\varphi(w,x)\| \to 0$ *as* $\varepsilon \to 0$ $\varphi(w,\cdot) \in C^3(\mathbf{R})$.

PROOF.– The statement of Lemma 3.2 becomes obvious if we use the definition of the generator of the Markov process and the form of the respective generators of the processes $\eta^\varepsilon(t)$ and $x(t/\varepsilon)$.

The truncated operator has the form

$$\Gamma_0^\varepsilon(x)\varphi(w) = \varepsilon^{-1}\mathbf{Q}\varphi(w,x) + \Gamma_1(x)\varphi(w,x). \qquad [3.6]$$

LEMMA 3.3.– *The solution of the singular perturbation problem for the truncated operator [3.6] on test functions*

$$\varphi^\varepsilon(w,x) = \varphi(w) + \varepsilon\varphi_1(w,x)$$

is implemented by the relation

$$\Gamma_0^\varepsilon(x)\varphi^\varepsilon(w,x) = \Gamma\varphi(w) + \varepsilon\theta_\eta^\varepsilon(x)\varphi(w), \qquad [3.7]$$

where the remainder term $\theta_\eta^\varepsilon(x)\varphi(w)$ *is uniformly bounded in* x.

The limit operator is defined by the formula

$$\Gamma = \Pi\Gamma_1(X)\Pi \qquad [3.8]$$

PROOF.– For equality [3.7] to hold, the coefficients of identical degrees ε on the left and on the right should coincide. Let us calculate

$$\begin{aligned}\Gamma_0^\varepsilon(x)\varphi^\varepsilon(w,x) &= \varepsilon^{-1}\mathbf{Q}\varphi(w) + [\mathbf{Q}\varphi_1(w,x) + \Gamma_1(x)\varphi(w)]\\ &\quad + \varepsilon\Gamma_1(x)\varphi_1(w,x),\end{aligned}$$

$$\mathbf{Q}\varphi(w) = 0 \Leftrightarrow \varphi(w) \in N_{\mathbf{Q}}.$$

From here, we can see that $\varphi(w)$ does not depend on x.

Consider the equation that defines the limit operator $\Gamma(x)$:

$$\mathbf{Q}\varphi_1(w,x) + \Gamma_1(x)\varphi(w) = \Gamma(x)\varphi(w).$$

Rewrite it as

$$\mathbf{Q}\varphi_1(w,x) = [\Gamma(x) - \Gamma_1(x)\varphi(w)]\varphi(w).$$

The resolvability condition for the last equation gives the limit operator in the form [3.8]. Then

$$\varphi_1(w,x) = R_0[\Gamma_1(x) - \Gamma]\varphi(w). \qquad [3.9]$$

Using [3.9], we can reduce the other terms of the expansion to the form

$$\varepsilon\Gamma_1(x)\varphi_1(w,x) = \varepsilon[\Gamma_1(x)R_0[\Gamma_1(x) - \Gamma]]\varphi(w) = \varepsilon\theta_\eta^\varepsilon(x)\varphi(w).$$

The boundedness of $\theta_\eta^\varepsilon(x)\varphi(w)$ follows from the form of the operators $\Gamma_1(x)$ and R_0.

The proof of Theorem 3.1 is completed using Lemma 3 and Theorem 6.3 from Korolyuk and Limnios (2005).

Consider the asymptotic properties of the original evolutionary system [4.1].

THEOREM 3.2.– *If conditions P1 – P4 are satisfied, the weak convergence*

$$u^\varepsilon(t) \to \hat{u}(t), \ \ \varepsilon \to 0$$

takes place. The limiting process $\hat{u}(t)$ *is defined by the generator*

$$\mathbf{L}\varphi(w) = \hat{C}(u)\varphi'(w) + \Gamma\varphi(w), \qquad [3.10]$$

where $\hat{C}(u) = \int\limits_X \pi(dx)C(u,x)$.

NOTE 3.1.– *The limiting process* $\hat{u}(t)$ *has two components. The deterministic drift is defined be the solution of the differential equation*

$$d\hat{u}_d(t) = [\hat{C}(\hat{u}_d(t)) + \tilde{a}]dt, \qquad [3.11]$$

where the additional term $\tilde{a}$ *appears due to the accumulation (with the normalized time* t/ε, $t \to 0$*) of small jumps of the impulse process that happen with a probability close to 1. The second component is the rare big jumps that take place with nearly zero probability and are defined in terms of averaged measure of jumps* $\tilde{\Gamma}_0(dv)$ *by the generator*

$$\Gamma_j\varphi(w) = \int\limits_R [\varphi(w+v) - \varphi(v) - v\varphi'(w)]\tilde{\Gamma}_0(dv).$$

NOTE 3.2.– *The limiting process* $\hat{u}(t)$ *is purely deterministic and defined by equation [3.11] in the case of zero average measure of jumps* $\tilde{\Gamma}_0(dv)$*. For example, if all the moments of order three and above for the family of processes with independent increments* $\eta^\varepsilon(t,x)$, $t \geq 0$, $x \in X$, *are equal to zero, or if the balance condition is satisfied*

$$\tilde{\Gamma}_0(v) = \Pi\Gamma_0(v,x) = \int\limits_X \pi(dx)\Gamma_0(v,x) = 0.$$

PROOF.–

Before directly proving Theorem 3.2, consider some propositions.

LEMMA 3.4.– *The generator of the two-component Markov process $u^\varepsilon(t), x(t/\varepsilon), t \geq 0$, has the representation*

$$\mathbf{L}^\varepsilon(x)\varphi(w,x) = \varepsilon^{-1}\mathbf{Q}\varphi(w,x) + \Gamma^\varepsilon(x)\varphi(w,x) \\ + \mathbf{C}(x)\varphi(w,x) + \theta^\varepsilon_w\varphi(w,x), \quad [3.12]$$

where $\Gamma^\varepsilon(x)$ is the generator of the family of processes with independent increments [3.3],

$$\mathbf{C}(x)\varphi(w,x) = C(u,x)\varphi'_w(w,x)$$

The remainder term $\|\theta^\varepsilon_w(x)\varphi(w,x)\| \to 0$ as $\varepsilon \to 0$.

PROOF.– Proof of the similar lemma is presented in Chabanyuk *et al.* (2019).

LEMMA 3.5.– *The generator $\mathbf{L}^\varepsilon(x)$ has the asymptotic representation*

$$\mathbf{L}^\varepsilon(x)\varphi(w,x) = \varepsilon^{-1}\mathbf{Q}\varphi(w,x) \\ + \Gamma_1\varphi(w,x) + \mathbf{C}(x)\varphi(w,x) + \hat{\theta}^\varepsilon_w\varphi(w,x), \quad [3.13]$$

where $\hat{\theta}^\varepsilon_w\varphi(w,x) = \gamma^\varepsilon + \theta^\varepsilon_w\varphi(w,x)$, and $\Gamma_1(x)$ is defined in Lemma 3.1.

The remainder term

$$\|\hat{\theta}^\varepsilon_w(x)\varphi(w,x)\| \to 0$$

when $\varepsilon \to 0$.

Proof uses the representation of the operator [3.5] and the results of Lemma 3.4.

The truncated operator has the form

$$\mathbf{L}^\varepsilon_0(x)\varphi = \varepsilon^{-1}\mathbf{Q}\varphi + \Gamma_1(x)\varphi + \mathbf{C}(x)\varphi. \quad [3.14]$$

LEMMA 3.6.– *Solution of the singular perturbation problem for the truncated operator [3.14] on test functions*

$$\varepsilon^(w,x) = \varphi(w) + \varepsilon\varphi_1(w,x)$$

is carried out by the relation

$$\mathbf{L}^\varepsilon_0(x)\varphi^\varepsilon(w,x) = \mathbf{L}\varphi(w) + \varepsilon\theta^\varepsilon_w(x)\varphi(w), \quad [3.15]$$

where the remainder term $\theta^\varepsilon_w(x)$ is uniformly bounded with respect to x.

The limit operator $\mathbf{L}$ is defined by the formula

$$\mathbf{L} = \Pi[C(x) + \Gamma_1(x)]\Pi. \quad [3.16]$$

For equality [3.15] to hold, the coefficients of identical degrees ε on the left and on the right should coincide. To this end, we will calculate

$$\mathbf{L}_0^\varepsilon(x)\varphi^\varepsilon(w,x) = \varepsilon^{-1}\mathbf{Q}(x)\varphi(u,w) + [\mathbf{Q}\varphi_1(w,x) + \Gamma_1(x)\varphi(w) + C(x)\varphi(w)] + \varepsilon\Gamma_1(x)\varphi_1(w,x).$$

Since

$$\mathbf{Q}\varphi(w) = 0 \Leftrightarrow \varphi(w) \in N_Q,$$

it is obvious that $\varphi(w)$ does not depend on x.

Let us rewrite the equation

$$\mathbf{Q}\varphi_1(w,x) + \Gamma_1(x)\varphi(w) + \mathbf{C}(x)\varphi(w) = \mathbf{L}\varphi(w)$$

as

$$\mathbf{Q}\varphi_1(w,x) = [\mathbf{L} - \Gamma_1(x) - \mathbf{C}(x)]\varphi(w).$$

The condition of resolvability of the last equation gives the limit operator $\mathbf{L}$ in the form [3.16].

The proof of the theorem is completed, which is similar to the proof of Theorem 6.3 from Korolyuk and Limnios (2005).

3.2. Asymptotic dissipativity of stochastic processes with impulse perturbation in the Poisson approximation scheme

In this section, we investigate the following question: how does the behavior of the limiting process depend on the prelimiting normalization of the stochastic evolutionary system in an ergodic Markov environment? We investigate the conditions of asymptotic dissipativity of the prelimiting system in the Poisson approximation scheme.

We consider the stochastic evolutionary system [3.1] and assume that the conditions of point 3.1 hold.

THEOREM 3.3.– *Let there exist the Lyapunov function $V(u) \in C^3(\mathbb{R}^d)$ of the system*

$$\frac{du}{dt} = \alpha(u), \qquad [3.17]$$

where $\alpha(u) = \hat{C}(u) + \hat{a}$, which satisfies the following conditions:

C1: $|\Gamma_u^1(x)R_0\hat{L}V(u)| < M_1V(u),\ M_1 > 0;$
C2: $|\Gamma_u^1(x)R_0\Gamma_u^1(x)V(u)| < M_2V(u),\ M_2 > 0;$
C3: $|\Gamma_u^1(x)R_0(C)(x)V(u)| < M_3V(u),\ M_3 > 0;$
C4: $|\mathbf{C}(x)R_0\hat{L}V(u)| < M_4V(u),\ M_4 > 0;$
C5: $|\mathbf{C}(x)R_0\Gamma_u^1(x)V(u)| < M_5V(u),\ M_5 > 0;$
C6: $|\mathbf{C}(x)R_0\mathbf{C}(x)V(u)| < M_6V(u),\ M_6 > 0.$

Suppose that inequalities

$$\alpha(u)V'(u) < -c_1V(u), \qquad [3.18]$$

$$|\int_X v^2\Gamma_0(dv,x)| < c_3(x), \qquad [3.19]$$

where $c_1 > 0$, $c_2 > 0$ *and* $\hat{c}_3 = \int_X \pi(dx)c_3(x) > 0$, *hold true.*

Then, system [3.1] is asymptotic dissipative.

Before directly proving Theorem 3.3, consider some propositions.

LEMMA 3.7.– *The generator of the three-component Markov process* $u^\varepsilon(t)$, $x(t/\varepsilon,\eta^\varepsilon(t,x))$, $t \geq 0$, *has the form*

$$\mathbf{L}^\varepsilon(x)\varphi(u,w,x) = \varepsilon^{-1}\mathbf{Q}\varphi(u,w,x) + \Gamma_w^\varepsilon(x)\varphi(u,w,x) + \mathbf{C}(x)\varphi(u,w,x) + \Gamma_u^\varepsilon(x)\varphi(u,w,x), \qquad [3.20]$$

where $\Gamma_w^\varepsilon(x)$ *is the generator of the family of processes with independent increments [3.4], acting at the variable* w, *and* $\Gamma_u^\varepsilon(x)$ *is equivalent to* $\Gamma_w^\varepsilon(x)$, *which is the generator of the family of processes with independent increments [3.4], acting at the variable* u.

PROOF.– The generator of the Markov process on perturbed test functions is defined from the relation

$$\mathbf{L}^\varepsilon(x)\varphi(u,w,x) = \lim_{\Delta\to 0}\frac{1}{\Delta}\mathbb{E}[\varphi(u_{t+\Delta}^\varepsilon, w_{t+\Delta}^\varepsilon, x_{t+\Delta}^\varepsilon) - \varphi(u,w,x)|u^\varepsilon(t) = u,\ \eta^\varepsilon(t) = w,\ x(t/\varepsilon) = x].$$

In the conditional expectation, add and subtract a value $\varphi(u, w_{t+\Delta}^\varepsilon, x_{t+\Delta}^\varepsilon)$:

$$\mathbb{E}[\varphi(u_{t+\Delta}^\varepsilon, w_{t+\Delta}^\varepsilon, x_{t+\Delta}^\varepsilon) - \varphi(u,w,x)|u^\varepsilon(t) = u, \eta^\varepsilon(t) = w, x(t/\varepsilon) = x]$$
$$= \mathbb{E}[\varphi(u_{t+\Delta}^\varepsilon, w_{t+\Delta}^\varepsilon, x_{t+\Delta}^\varepsilon) - \varphi(u, w_{t+\Delta}^\varepsilon, x_{t+\Delta}^\varepsilon)]$$
$$= \mathbb{E}[\varphi(u, w_{t+\Delta}^\varepsilon, x_{t+\Delta}^\varepsilon) - \varphi(u,w,x)].$$

The expansion of $u^\varepsilon_{t+\Delta}$ has the form

$$u^\varepsilon_{t+\Delta} = u + C(u,x)\Delta + \Delta w + o(\Delta).$$

The expression obtained will be substituted into the first addition of the conditional expectation:

$$\begin{aligned}&\mathbb{E}[\varphi(u^\varepsilon_{t+\Delta}, w^\varepsilon_{t+\Delta}, x^\varepsilon_{t+\Delta}) - \varphi(u, w^\varepsilon_{t+\Delta}, x^\varepsilon_{t+\Delta})]\\ &= \mathbb{E}[\varphi(u + C(u,x)\Delta + \Delta w + o(\Delta), w^\varepsilon_{t+\Delta}, x^\varepsilon_{t+\Delta}) - \varphi(u, w^\varepsilon_{t+\Delta}, x^\varepsilon_{t+\Delta})]\\ &= \mathbb{E}[\varphi(z + \Delta z, w^\varepsilon_{t+\Delta}, x^\varepsilon_{t+\Delta}) - \varphi(u, w^\varepsilon_{t+\Delta}, x^\varepsilon_{t+\Delta})],\end{aligned}$$

where

$$z = u + C(u,x)\Delta + o(\Delta).$$

Then, we add and subtract $\varphi(z, w^\varepsilon_{t+\Delta}, x^\varepsilon_{t+\Delta})$ from the expression obtained:

$$\begin{aligned}&\mathbb{E}[\varphi(u^\varepsilon_{t+\Delta}, w^\varepsilon_{t+\Delta}, x^\varepsilon_{t+\Delta}) - \varphi(u, w^\varepsilon_{t+\Delta}, x^\varepsilon_{t+\Delta})]\\ &= \mathbb{E}[\varphi(z + \Delta w, w^\varepsilon_{t+\Delta}, x^\varepsilon_{t+\Delta}) - \varphi(z, w^\varepsilon_{t+\Delta}, x^\varepsilon_{t+\Delta})]\\ &\quad + \mathbb{E}[\varphi(z, w^\varepsilon_{t+\Delta}, x^\varepsilon_{t+\Delta}) - \varphi(u, w^\varepsilon_{t+\Delta}, x^\varepsilon_{t+\Delta})].\end{aligned}$$

Since the generator $\Gamma^\varepsilon_u(x)$ has the form

$$\Gamma^\varepsilon_u(x)\varphi(u,w,x) = \lim_{\Delta\to 0}\frac{1}{\Delta}(\mathbb{E}[\varphi(u + \Delta u, w, x) - \varphi(u,w,x)]),$$

for the limit of the first addition, we obtain

$$\lim_{\Delta\to 0}\frac{1}{\Delta}\mathbb{E}[\varphi(z + \Delta w, w^\varepsilon_{t+\Delta}, x^\varepsilon_{t+\Delta}) - \varphi(z, w^\varepsilon_{t+\Delta}, x^\varepsilon_{t+\Delta})] = \Gamma^\varepsilon_u(x)\varphi(u,w,x).$$

Let us expand $\varphi(z, w^\varepsilon_{t+\Delta}, x^\varepsilon_{t+\Delta})$ according to the Taylor formula

$$\begin{aligned}&\varphi(u + C(u,x)\Delta + o(\Delta), w^\varepsilon_{t+\Delta}, x^\varepsilon_{t+\Delta})\\ &= \varphi(u, w^\varepsilon_{t+\Delta}, x^\varepsilon_{t+\Delta}) + \varphi', w^\varepsilon_{t+\Delta}, x^\varepsilon_{t+\Delta})(C(u,x)\Delta + o(\Delta)) + o(\Delta).\end{aligned}$$

Substituting into the expression $\mathbb{E}[\varphi(z, w^\varepsilon_{t+\Delta}, x^\varepsilon_{t+\Delta}) - \varphi(u, w^\varepsilon_{t+\Delta}, x^\varepsilon_{t+\Delta})]$ the expansion obtained, we have

$$\begin{aligned}&\lim_{\Delta\to 0}\frac{1}{\Delta}\mathbb{E}[\varphi(z, w^\varepsilon_{t+\Delta}, x^\varepsilon_{t+\Delta}) - \varphi(u, w^\varepsilon_{t+\Delta}, x^\varepsilon_{t+\Delta})]\\ &= \lim_{\Delta\to 0}\frac{1}{\Delta}\mathbb{E}[\varphi(u + C(u,x)\Delta + o(\Delta), w^\varepsilon_{t+\Delta}, x^\varepsilon_{t+\Delta}) - \varphi(u, w^\varepsilon_{t+\Delta}, x^\varepsilon_{t+\Delta})]\end{aligned}$$

$$= \lim_{\Delta \to 0} \frac{1}{\Delta} \mathbb{E}[\varphi(u, w^\varepsilon_{t+\Delta}, x^\varepsilon_{t+\Delta}) + \varphi'(u, w^\varepsilon_{t+\Delta}, x^\varepsilon_{t+\Delta})(C(u,x)\Delta + o(\Delta))$$

$$+o(\Delta) - \varphi(u, w^\varepsilon_{t+\Delta}, x^\varepsilon_{t+\Delta})]$$

$$= \lim_{\Delta \to 0} \frac{1}{\Delta} \mathbb{E}[\varphi'(u, w^\varepsilon_{t+\Delta}, x^\varepsilon_{t+\Delta})(C(u,x)\Delta + o(\Delta)) + o(\Delta)]$$

$$= C(u,x)\varphi'(u,w,x).$$

Similarly, from the relation of the generator $\Gamma^\varepsilon_w(x)$ and the obvious relation of the Markov process generator

$$\lim_{\Delta \to 0} \frac{1}{\Delta} \mathbb{E}[\varphi(w, x^\varepsilon_{t+\Delta}) - \varphi(w,x)] = \varepsilon^{-1}\mathbf{Q}\varphi(w,x),$$

we have

$$\lim_{\Delta \to 0} \frac{1}{\Delta} \mathbb{E}[\varphi(u, w^\varepsilon_{t+\Delta}), x^\varepsilon_{t+\Delta}) - \varphi(u,w,x)]$$

$$= \lim_{\Delta \to 0} \frac{1}{\Delta} \mathbb{E}[\varphi(u, w^\varepsilon_{t+\Delta}), x^\varepsilon_{t+\Delta}) - \varphi(u,w,x^\varepsilon_{t+\Delta}))$$

$$+\varphi(u,w,x^\varepsilon_{t+\Delta})) - \varphi(u,w,x)]$$

$$= \lim_{\Delta \to 0} \frac{1}{\Delta} \mathbb{E}[\varphi(u, w^\varepsilon_{t+\Delta}), x^\varepsilon_{t+\Delta}) - \varphi(u,w,x^\varepsilon_{t+\Delta})]$$

$$+ \lim_{\Delta \to 0} \frac{1}{\Delta} \mathbb{E}[\varphi(u,w,x^\varepsilon_{t+\Delta}) - \varphi(u,w,x)]$$

$$= \Gamma^\varepsilon_w(x)\varphi(u,w,x) + \varepsilon^{-1}\mathbf{Q}\varphi(u,w,x),$$

and, finally, we can see that $\mathbf{L}^\varepsilon(x)$ has the form [3.20].

LEMMA 3.8.– *The generator [3.20] has the asymptotic representation*

$$\mathbf{L}^\varepsilon(x)\varphi(u,w,x) = \varepsilon^{-1}\mathbf{Q}\varphi(u,w,x) + \Gamma^1_u(x)\varphi(u,w,x)$$
$$+ C(u)\varphi(u,w,x) + \Gamma^1_w(x)\varphi(u,w,x) + \gamma^\varepsilon(x)\varphi(w,x)$$

where

$$C(u)\varphi(u) = C(u,x)\varphi'(u),$$

and the remainder term

$$\|\gamma^\varepsilon(x)\varphi(u,w,x)\| \to 0, \ \varepsilon \to 0,$$

$$\varphi(u,w,\cdot) \in C^3((R)).$$

The truncated generator has the form

$$\mathbf{L}_0^\varepsilon(x)\varphi(u,w,x) = \varepsilon^{-1}\mathbf{Q}\varphi(u,w,x) + \Gamma_u^1(x)\varphi(u,w,x) + C(u)\varphi(u,w,x) + \Gamma_w^1(x)\varphi(u,w,x). \qquad [3.21]$$

LEMMA 3.9.– *The solution of the singular perturbation problem for the truncated generator [3.21] on perturbed test functions*

$$\varphi^\varepsilon(u,w,x) = \varphi(u,w) + \varepsilon\varphi_1(u,w,x) \qquad [3.22]$$

is determined by the equality

$$L_0^\varepsilon(x)\varphi^\varepsilon(u,w,x) = \hat{L}\varphi(u,w) + \varepsilon\theta^\varepsilon(x)\varphi(u,w),$$

where

$$\begin{aligned}\theta^\varepsilon(x) = {} & \Gamma_u^1(x)R_0\hat{L} - \Gamma_u^1(x)R_0\Gamma_u^1(x) \\ & - \Gamma_u^1(x)R_0\mathbf{C}(x) - \Gamma_u^1(x)R_0\Gamma_w^1(x) \\ & + \mathbf{C}(x)R_0\hat{L} - \mathbf{C}(x)R_0\Gamma_u^1(x) \\ & - \mathbf{C}(x)R_0\mathbf{C}(x) - \mathbf{C}(x)R_0\Gamma_w^1(x) \\ & + \Gamma_w^1(x)R_0\hat{L} - \Gamma_w^1(x)R_0\Gamma_u^1(x) \\ & - \Gamma_w^1(x)R_0\mathbf{C}(x) - \Gamma_w^1(x)R_0\Gamma_w^1(x).\end{aligned}$$

PROOF.– Let us substitute [3.22] into [3.21].

$$\begin{aligned}\mathbf{L}_0^\varepsilon(x)\varphi(u,w,x) = {} & \varepsilon^{-1}\mathbf{Q}[\varphi(u,w) + \varepsilon\varphi_1(u,w,x)] \\ & + \Gamma_u^1(x)[\varphi(u,w) + \varepsilon\varphi_1(u,w,x)] \\ & + \mathbf{C}(x)[\varphi(u,w) + \varepsilon\varphi_1(u,w,x)] \\ & + \Gamma_w^1[\varphi(u,w) + \varepsilon\varphi_1(u,w,x)] \\ = {} & \varepsilon^{-1}\mathbf{Q}\varphi(u,w) + [\mathbf{Q}\varphi_1(u,w,x) + \Gamma_u^1(x)\varphi(u,w) \\ & + \mathbf{C}(x)\varphi(u,v) + \Gamma_w^1(x)\varphi(u,w)] + \varepsilon[\Gamma_u^1(x)\varphi_1(u,w,x) \\ & + \mathbf{C}(x)\varphi_1(u,w,x) + \Gamma_w^1(x)\varphi_1(u,w,x)].\end{aligned}$$

For the existing limit operator $\hat{L}\varphi(u,w)$, it is necessary that, at $\varepsilon \to 0$, it satisfies the condition

$$\mathbf{Q}\varphi(u,w) = 0,$$

i.e. the function must belong to the zero-subspace of the operator $\mathbf{Q}$.

Then

$$\hat{L}\varphi(u,w) = \mathbf{Q}\varphi_1(u,w,x) + \Gamma_u^1(x)\varphi(u,w) \\ + \mathbf{C}(x)\varphi(u,w) + \Gamma_w^1(x)\varphi(u,w),$$

where

$$\mathbf{Q}\varphi_1(u,w,x) = [\hat{L} - \Gamma_u^1(x) - \mathbf{C}(x) - \Gamma_w^1(x)]\varphi(u,w).$$

From the condition for resolvability to the last equation, we have

$$\Pi\mathbf{Q}\Pi\varphi_1(u,w,x) = 0 = \Pi[\hat{L} - \Gamma_u^1(x) - \mathbf{C}(x) - \Gamma_w^1(x)]\Pi\varphi(u,w).$$

Thus

$$\hat{L}\varphi(u,w) = \Pi\Gamma_u^1(x)\varphi(u,w) + \Pi\mathbf{C}(x)\varphi(u,w) + \Pi\Gamma_w^1(x)\varphi(u,w),$$

and

$$\varphi_1(u,w,x) = R_0[\hat{L} - \Gamma_u^1(x) - \mathbf{C}(x) - \Gamma_w^1(x)]\varphi(u,w).$$

We obtain the relation for the last equation:

$$\begin{aligned}
&\varepsilon[\Gamma_u^1(x)\varphi_1(u,w,x) + \mathbf{C}(x)\varphi_1(u,w,x) + \Gamma_w^1(x)\varphi_1(u,w,x)] \\
&= \varepsilon[\Gamma_u^1(x)R_0[\hat{L} - \Gamma_u^1(x) - \mathbf{C}(x) - \Gamma_w^1(x)] \\
&\quad +\mathbf{C}[\hat{L} - \Gamma_u^1(x) - \mathbf{C}(x) - \Gamma_w^1(x)] \\
&\quad +\Gamma_w^1(x)R_0[\hat{L} - \Gamma_u^1(x) - \mathbf{C}(x) - \Gamma_w^1(x)]]\varphi(u,w).
\end{aligned}$$

PROOF OF THEOREM 3.3.– Under conditions C1 – C6, we have the boundedness of the remainder term [3.23]:

$$\begin{aligned}
&\|\theta^\varepsilon(x)V(u)\| \\
&= |\Gamma_u^1(x)R_0\hat{L}V(u) - \Gamma_u^1(x)R_0\Gamma_u^1(x)V(u) \\
&\quad -\Gamma_u^1(x)R_0\mathbf{C}(x)V(u) - \Gamma_u^1(x)R_0\Gamma_w^1(x)V(u) \\
&\quad +\mathbf{C}(x)R_0\hat{L}V(u) - \mathbf{C}(x)R_0\Gamma_u^1(x)V(u) \\
&\quad -\mathbf{C}(x)R_0\mathbf{C}(x)V(u) - \mathbf{C}(x)R_0\Gamma_w^1(x)V(u) \\
&\quad +\Gamma_w^1(x)R_0\hat{L}V(u) - \Gamma_w^1(x)R_0\Gamma_u^1(x)V(u) \\
&\quad -\Gamma_w^1(x)R_0\mathbf{C}(x)V(u) - \Gamma_w^1(x)R_0\Gamma_w^1(x)V(u)| \\
&\quad \le M_1V(u) + M_2V(u) + M_3V(u) + M_4V(u) + M_5V(u) + M_6V(u),
\end{aligned}$$

where

$$\|\theta^{\varepsilon}(x)V(u)\| \leq MV(u), \qquad [3.23]$$

and $M = M_1 + M_2 + M_3 + M_4 + M_5 + M_6$.

From Lemma 3.9, the relation [3.23] and the Korolyuk model theorem, we obtain the weak convergence

$$(u^{\varepsilon}(t), \eta^{\varepsilon}(t)) \Rightarrow (u(t), \eta(t)), \ \varepsilon \to 0.$$

Furthermore, let $\frac{d^{[3.1]}V(u)}{du}$ be a derivative of the Lyapunov function, which is calculated along the trajectory of system [3.1]. Since the Lyapunov function must satisfy the Lipschitz condition

$$|V(u_2) - V(u_1)| < K|u_2 - u_1|,$$

where K is a constant, then

$$\frac{d^{[3.1]}V(u)}{du} \leq \frac{dV(u)}{du} + K[\hat{a} + \int\limits_{R} v^2\Gamma_0(dv)]dt,$$

where $\frac{dV(u)}{du}$ is a derivative of the Lyapunov function, which is calculated along the trajectory of deterministic system [3.17],

$$\Gamma_0(dv) = \int\limits_{X} \pi(dx)\Gamma_0(dv, x).$$

According to conditions [3.18] and [3.19], we obtain

$$\frac{d^{[3.1]}V(u)}{du} \leq -c_1 V(u) + K[c_2 + \hat{c}_3]|dt|.$$

Thus,

$$V(u) \leq V(u_0)\exp\{-c_1\}t + K[c_2 + \hat{c}_3] \int\limits_{0}^{t} \exp\{-c_1(t-s)\}ds.$$

From the last relation, we have estimation

$$\mathbb{P}\{|u(t)| > R\} \leq \frac{V(u)}{\inf\limits_{u \in R^d} V(u)}, \ R \to \infty,$$

Thus, system [3.1] is dissipative, and from the Korolyuk model theorem and dissipativity of the limiting process, we conclude that system [3.1] is asymptotic dissipative.

3.3. Double merging of the phase space for differential equations with small stochastic supplements under Poisson approximation conditions

In this section, we consider the case when system perturbation is determined by the jumping process under the Poisson approximation scheme. In particular, we are interested in double merging of the phase space for such evolutionary models.

Let us consider an evolutionary system in ergodic Markov environment that has the following form of the stochastic differential equation:

$$du^\varepsilon(t) = C(u^\varepsilon(t), x(t/\varepsilon^2))dt + d\eta^\varepsilon(t),\ \ u^\varepsilon(t) \in \mathbb{R}, \qquad [3.24]$$

where the Markov process $x^\varepsilon(t), t \geq 0$ is determined in a standard phase space (E, E) with splitting

$$E = \bigcup_{k=1}^{N} E_k,\ \ E_k \cap E_{k'} = \emptyset,\ \ k \neq k'$$

at a series scheme with a small parameter of a series $\varepsilon \to 0$ when $\varepsilon > 0$.

The Markov kernel has the form

$$Q^\varepsilon(x, B, t) = P^\varepsilon(x, B)[1 - \exp\{-q(x)t\}],\ \ x \in E,\ \ B \in E,\ \ t \geq 0.$$

Let conditions ME1 – ME4 from the previous section hold true.

The impulse perturbation process $\eta^\varepsilon(t)$, $t \geq 0$ under the Poisson approximation scheme is given by the relation

$$\eta^\varepsilon(t) = \int_0^t \eta^\varepsilon(ds, x(s/\varepsilon^3)), \qquad [3.25]$$

where a set of processes with independent increments $\eta^\varepsilon(t, x)$, $t \geq 0$, $x \in X$, is determined by the generators

$$\Gamma^\epsilon(x)\varphi(w) = \varepsilon^{-2} \int_R (\varphi(w + v) - \varphi(w))\Gamma^\varepsilon(dv, x),\ \ x \in X \qquad [3.26]$$

and satisfies the Poisson approximation conditions

P1. Approximation of averages

$$\int_R v\Gamma^\varepsilon(dv, x) = \varepsilon(a(x) + \theta_a(x)),\ \ \theta_a(x) \to 0,\ \ \varepsilon \to 0,$$

and

$$\int_R v^2 \Gamma^\varepsilon(dv, x) = \varepsilon(b(x) + \theta_b(x)), \;\; \theta_b(x) \to 0, \;\; \varepsilon \to 0;$$

P2. Condition for the distribution function

$$\int_R g(v)\Gamma^\varepsilon(dv, x) = \varepsilon(\Gamma_g(x) + \theta_g(x)), \;\; \theta_g(x) \to 0, \;\; \varepsilon \to 0,$$

for each $g(v) \in C^2(\mathbb{R})$ (space of real values bounded functions $g(v)/|v|^2 \to 0$, $|v| \to 0$). The measure $\Gamma_g(x)$ is bounded for each $g(v) \in C^2(\mathbb{R})$ and is determined by the relation

$$\Gamma_g(x) = \int_R g(v)\Gamma_0(dv, x), \;\; g(v) \in C^2(\mathbb{R});$$

P3. Uniform quadratic integrability

$$\sup \lim_{c \to \infty} \int_{|v|>c} v^2 \Gamma_0(dv, x) = 0;$$

P4. The absence of a diffusion component

$$b(x) = \int_R v^2 \Gamma_0(dv, x).$$

Let us denote:

$$\Gamma_1(x)\varphi(w) = a(x)\varphi'(w) + \int_R [\varphi(w+v) - \varphi(v) - v\varphi'(w)]\Gamma_0(dv, x).$$

First, we investigate the asymptotic properties of the perturbation process.

THEOREM 3.4.– *P1 – P4 the weak convergence*

$$\eta^\varepsilon(t) \to \eta^0(t), \;\; \varepsilon \to 0.$$

holds true.

The limit process $\eta^0(t)$ *is determined by the generator*

$$\hat{\hat{\Gamma}}\varphi(w) = \hat{\Pi}\hat{\Gamma}_1(x)\varphi(w) = \hat{a}\varphi'(w) + \int_R [\varphi(w+v) - \varphi(w) - v\varphi'(w)]\hat{\hat{\Gamma}}_0(dv),$$

where

$$\hat{\hat{a}} = \sum_{k \in \hat{E}} \hat{\pi}_k \int_{\hat{E}_k} \pi(dx)(a_(x)),$$

$$\hat{\hat{\Gamma}}_0(v) = \sum_{k \in \hat{E}} \hat{\pi}_k \int_{\hat{E}_k} \pi(dx)\Gamma_0(v, x),$$

which is a process with independent increments, deterministic shift and Poisson jump component.

PROOF.– We begin by obtaining some propositions (Samoilenko and Nikitin 2019).

LEMMA 3.10.– *Generators of processes with independent increments* $\eta^\varepsilon(t, x)$, $t \geq 0$, $x \in X$, *on test functions* $\varphi(w) \in C^2(\mathbf{R})$ *under Poisson approximation conditions P1 – P4 have the asymptotic representation*

$$\mathbf{\Gamma}^\varepsilon(x)\varphi(w) = \Gamma_1(x)\varphi(w) + \gamma^\varepsilon(x)\varphi(w), \qquad [3.27]$$

where

$$\Gamma_1(x)\varphi(w) = a(x)\varphi'(w) + \int_R [\varphi(w + v) - \varphi(v) - v\varphi'(w)]\Gamma_0(dv, x),$$

and the remainder term $\|\gamma^\varepsilon(x)\varphi(w)\| \to 0$ *when* $\varepsilon \to 0$, $\varphi(w, \cdot) \in C^2(\mathbb{R})$.

PROOF.– Using a Taylor series of functions $\Gamma_1(x)\varphi(w)$, we can transform the generator [3.26]:

$$\begin{aligned}
\mathbf{\Gamma}^\varepsilon(x)\varphi(w) &= \varepsilon^{-1} \int_R (\varphi(w + v) - \varphi(v))\Gamma^\varepsilon(dv, x) \\
&= \varepsilon^{-1} \int_R (\varphi(w + v) - \varphi(v) - v\varphi'(v) - \frac{1}{2} v^2 \varphi''(w))\Gamma^\varepsilon(dv, x) \\
&\quad + \varepsilon^{-1} \int_R (v\varphi'(w)\Gamma^\varepsilon(dv, x) + \frac{1}{2} v^2 \varepsilon^{-1} \int_R v^2 \varphi''(w)\Gamma^\varepsilon(dv, x) \\
&= \int_R (\varphi(u + v) - \varphi(v) - v\varphi'(w) - \frac{1}{2} v^2 \varphi''(w))\Gamma_0(dv, x) \\
&\quad + a(x)\varphi'(w) + \frac{1}{2} b(x)\varphi''(w) + \gamma^\varepsilon(x)\varphi(w) \\
&= \int_R (\varphi(u + v) - \varphi(v) - \varphi'(w))\Gamma_0(dv, x) \\
&\quad + a(x)\varphi'(w) + \gamma^\varepsilon(w)\varphi(w),
\end{aligned}$$

where the preemption equality follows from conditions P1 – P4. We note that the function

$$\varphi(w+v) - \varphi(w) - v\varphi'(w) - \frac{1}{2}v^2\varphi''(w) \in C^2(\mathbf{R}),$$

since it is bounded because $\varphi(w)$ and its derivatives are bounded and the relation

$$[\varphi(w+v) - \varphi(w) - v\varphi'(w) - \frac{1}{2}v^2\varphi''(w)]/|v^2| \to 0$$

holds true when $v \to 0$.

Remembering $\gamma^\varepsilon(w)\varphi(w) = o(\varepsilon^2)$, $\varphi(w) \in C^2(\mathbb{R})$, we obtain [4.206].

LEMMA 3.11.– *The generator of a two-component Markov process* $(\eta^\varepsilon, x(t/\varepsilon^2))$, $t \geq 0$ *has the following form:*

$$\hat{\Gamma}^\varepsilon(x)\varphi(u,w,x) = \varepsilon^{-2}\mathbf{Q}^\varepsilon\varphi(u,w,x) + \gamma^\varepsilon(x)\varphi(u,w,x), \tag{3.28}$$

where $\Gamma_1(x)$ *is determined in Lemma 3.10, and the remainder term* $\|\gamma^\varepsilon(x)\varphi(u,w,x)\| \to 0$ *when* $\varepsilon \to 0$, $\varphi(u,w,\cdot) \in C^2(\mathbb{R})$.

PROOF.– The statement of the lemma follows from the definition of the Markov process generator and from the respective forms of the generators $\eta^\varepsilon(t,x)$ i $x(t/\varepsilon^2)$.

The truncated generator has the form

$$\mathbb{L}^\varepsilon\varphi(u,w,x) = \varepsilon^{-2}\mathbf{Q}\varphi(u,w,x) + \Gamma_1(x)\varphi(u,w,x). \tag{3.29}$$

LEMMA 3.12.– *The solution of singular perturbation problem for the truncated operator* [3.29] *on test functions*

$$\varphi^\varepsilon(u,w,x) = \varphi(u,w) + \varepsilon\varphi_1(u,w,x) + \varepsilon^2\varphi_2(u,w,x)$$

can be given by the relation

$$\Gamma_0^\varepsilon(x)\varphi^\varepsilon(u,x) = \hat{\hat{L}}\varphi(u) + \varepsilon\theta_\eta^\varepsilon(x)\varphi(u), \tag{3.30}$$

where the remainder term is uniformly bounded with respect to x.

The limit operator can be determined by the formula

$$\hat{\hat{L}} = \hat{\Pi}\hat{\Gamma}_1\hat{\Pi} \tag{3.31}$$

PROOF.– Let us calculate

$$\begin{aligned}&(\varepsilon^{-2}Q + \varepsilon^{-1}Q_1 + \Gamma_1)(\varphi(u) + \varepsilon\varphi_1 + \varepsilon^2\varphi_2)\\&= \varepsilon^{-2}Q\varphi + \varepsilon^{-1}(Q\varphi_1 + Q_1\varphi) + (Q\varphi_2 + Q_1\varphi_1 + \Gamma_1\varphi) + o(\varepsilon).\end{aligned}$$

We have three relations:

$$Q\varphi = 0; \qquad [3.32]$$

$$Q\varphi_1 + Q_1\varphi = 0; \qquad [3.33]$$

$$Q\varphi_2 + Q_1\varphi_1 + \Gamma_1\varphi = \hat{\hat{L}}\varphi. \qquad [3.34]$$

Furthermore, we find the form of $\hat{\hat{L}}$.

From [3.32] follows $\varphi \in N_Q$.

In [3.33], since $\varphi \in N_Q$, from the solvability condition, we obtain

$$\Pi Q_1 \Pi \varphi = 0.$$

Let us denote

$$\Pi Q_1 = \hat{Q}_1, \ \Pi\varphi = \hat{\varphi}.$$

Then

$$\hat{Q}_1\hat{\varphi} = 0,$$

where

$$\hat{\varphi} \in N_{\hat{Q}_1}.$$

Let us consider [3.34]: from the solvability condition for **Q**, we obtain

$$\Pi Q_1 \Pi \varphi_1 + \Pi\Gamma_1\Pi\varphi = \Pi\hat{\hat{L}}\Pi\varphi,$$

where

$$\hat{Q}_1\hat{\varphi}_1 + \hat{\Gamma}_1\hat{\varphi} = \hat{\hat{L}}\hat{\varphi}.$$

Furthermore, from the solvability condition for $\hat{\varphi}_2$, we obtain

$$\hat{\Pi}\hat{\Gamma}_1\hat{\Pi}\hat{\varphi} = \hat{\hat{L}}\hat{\hat{\varphi}},$$

where

$$\hat{\hat{L}} = \hat{\Pi}\hat{\Gamma}_1\hat{\Pi},$$

$$\hat{\varphi}_1 = \hat{R}_0[\hat{\Gamma}_1 - \hat{\hat{L}}]\hat{\varphi},$$

$$\hat{\varphi}_2 = R_0[Q_1\varphi_1 + \Gamma_1\varphi - \hat{\hat{L}}\varphi].$$

The boundedness of $\theta^{\varepsilon}_{\eta}(x)\varphi(w)$ follows from the form of the operators Γ_1 and R_0.

The proof of the theorem is completed using Lemma 3.12 and Theorem 6.3 in Korolyuk and Limnios (2005).

Furthermore, we investigate the asymptotic properties of the evolutionary system [2.27]

THEOREM 3.5.– *Under conditions P1 – P4, the weak convergence*

$$(u^\varepsilon(t), \eta^\varepsilon(t)) \Rightarrow (\hat{u}(t), \eta^0(t)), \ \varepsilon \to 0.$$

holds true.

The limit process $(\hat{u}(t), \eta^0(t))$ *can be determined by the generator*

$$\mathbf{L}\varphi(u,w) = \hat{\hat{C}}(u)\varphi'_u(u,w) + \hat{\hat{\Gamma}}^w\varphi(\cdot,w), \quad [3.35]$$

where

$$\hat{\hat{C}}(u) = \Pi\mathbf{C}(x) = \sum_{k\in\hat{E}} \hat{\pi}_k \int_{\hat{E}_k} \pi(dx)C(u,x);$$

and the generator $\hat{\hat{\Gamma}}^w$ *is defined in Theorem 3.4, acting at the variable* w.

NOTE 3.3.– *Weak convergences of processes* $u^\varepsilon(t) \Rightarrow \hat{u}(t)$, $\varepsilon \to 0$, *follow from the convergence of the respective generators under the assumption of the compactness of a prelimiting set of processes* $u^\varepsilon(t)$. *The respective theorems can be found, in particular, in Korolyuk and Limnios (2005).*

PROOF.–

LEMMA 3.13.– *The generator of a three-component Markov process* $(u^\varepsilon(t), \eta^\varepsilon(t), x^\varepsilon(t/\varepsilon^3))$, $t \geq 0$, *has the representation*

$$\mathbf{L}^\varepsilon(x)\varphi(u,w,x) = \varepsilon^{-2}Q^\varepsilon\varphi(u,w,x) + \varepsilon\Gamma^\varepsilon_u(x)\varphi(u,\cdot,x) + \Gamma^\varepsilon_w(x)\varphi(\cdot,w,x) +$$

$$+\mathbf{C}(x)\varphi(u,w,x) + \theta^\varepsilon_w(x)\varphi(u,w,x), \quad [3.36]$$

where $\Gamma^\varepsilon_\cdot(x)$ *is the IPP generator* [3.26],

$$\mathbf{C}(x)\varphi(u,w,x) = C(u,x)\varphi'_u(u,w,x).$$

The remainder term $\|\theta^\varepsilon_w(x)\varphi(u,w,x)\| \to 0$ *when* $\varepsilon \to 0$.

The proof can be found in Korolyuk and Limnios (2005).

LEMMA 3.14.– *The generator* $\mathbf{L}^\varepsilon(x)$ *in the case of the IPP has the asymptotic representation*

$$\mathbf{L}^\varepsilon(x)\varphi(u,w,x) = \varepsilon^{-2}Q^\varepsilon\varphi(u,w,x) + \varepsilon\Gamma^u_1(x)\varphi(u,w,x) + \Gamma^w_1(x)\varphi(u,w,x)$$

$$+ \mathbf{C}(x)\varphi(u,w,x) + \hat{\theta}^\varepsilon_w\varphi(u,w,x), \quad [3.37]$$

where

$$\hat{\theta}^\varepsilon_w(x) = \gamma^\varepsilon + \theta^\varepsilon_w(x),$$

$\Gamma_1^{\cdot}(x)$ *is defined in Lemma 3.10*

The remainder term $\|\hat{\theta}_w^{\varepsilon}(x)\varphi(u,w,x)\| \to 0$ *when* $\varepsilon \to 0$.

PROOF.– The proof follows from the representation of the generator [3.27] and Lemma 3.13.

The truncated operator has the form:

$$\mathbf{L}_0^{\varepsilon}(x)\varphi = \varepsilon^{-2}Q^{\varepsilon}\varphi + \Gamma_1^{w}(x)\varphi + \mathbf{C}(x)\varphi \qquad [3.38]$$

LEMMA 3.15.– *The solution of the singular perturbation problem for the truncated operator* [3.38] *on test functions*

$$\varphi^{\varepsilon}(w,x) = \varphi(w) + \varepsilon\varphi_1(w,x) + \varepsilon^2\varphi_2(w,x)$$

holds by the relation

$$\mathbf{L}_0^{\varepsilon}(x)\varphi^{\varepsilon}(w,x) = \mathbf{L}\varphi(w) + \varepsilon^3\theta_w^{\varepsilon}(x)\varphi(w), \qquad [3.39]$$

where the remainder term $\theta_w^{\varepsilon}(x)$ *is uniformly bounded at* x.

The limit operator $\mathbf{L}$ *is defined by the formula*

$$\mathbf{L} = \hat{\hat{C}} + \hat{\hat{\Gamma}}_1^{w}. \qquad [3.40]$$

PROOF.– In order to satisfy [3.39], it is necessary that the coefficients at the same degrees ε of the left and right be equal. For this purpose, we will calculate

$$(\varepsilon^{-2}Q + \varepsilon^{-1}Q_1 + \Gamma_1^{w} + \mathbf{C})(\varphi + \varepsilon\varphi_1 + \varepsilon^2\varphi_2)$$

$$= \varepsilon^{-2}Q\varphi + \varepsilon^{-1}(Q\varphi_1 + Q_1\varphi) + (Q\varphi_2 + Q_1\varphi_1 + \Gamma_1^{w}\varphi + \mathbf{C}\varphi) + o(\varepsilon).$$

and, again, we have three relations:

$$Q\varphi = 0; \qquad [3.41]$$

$$Q\varphi_1 + Q_1\varphi = 0; \qquad [3.42]$$

$$Q\varphi_2 + Q_1\varphi_1 + \Gamma_1^{w}\varphi = \hat{\hat{L}}\varphi. \qquad [3.43]$$

We define the form of $\hat{\hat{L}}$.

From [3.41], it follows that $\varphi \in N_Q$.

In [3.42], since $\varphi \in N_Q$, according to the solvability condition, we obtain

$$\Pi Q_1 \Pi\varphi = 0.$$

Let us denote

$$\Pi Q_1 = \hat{Q}_1, \ \Pi\varphi = \hat{\varphi}.$$

Then

$$\hat{Q}_1\hat{\varphi} = 0,$$

where

$$\hat{\varphi} \in N_{\hat{Q}_1}.$$

Let us consider [3.43]: from the solvability condition for Q, we have

$$\Pi Q_1 \Pi \varphi_1 + \Pi \Gamma_1^w \Pi \varphi + \Pi \mathbf{C} \Pi \varphi = \Pi \hat{\hat{L}} \Pi \varphi. \tag{3.44}$$

Furthermore

$$\hat{Q}_1\hat{\varphi}_1 + \hat{\Gamma}_1^w \hat{\varphi} + \hat{C}\hat{\varphi} = \hat{\hat{L}}\hat{\varphi}.$$

Again, from the solvability condition for $\hat{\varphi}_2$, we obtain

$$\hat{\Pi}\hat{C}\hat{\Pi}\hat{\varphi} + \hat{\Pi}\hat{\Gamma}_1^w\hat{\Pi}\hat{\varphi} = \hat{\hat{L}}\hat{\hat{\varphi}},$$

where

$$\hat{\hat{L}} = \hat{\Pi}\hat{C}\hat{\Pi} + \hat{\Pi}\hat{\Gamma}_1^w\hat{\Pi},$$

$$\hat{\varphi}_1 = \hat{R}_0[\hat{\hat{L}} - \hat{\Gamma}_1^w\varphi_1 - \hat{C}]\hat{\varphi},$$

$$\hat{\varphi}_2 = \hat{R}_0[\hat{\hat{L}}\varphi + Q_1\varphi_1 + \Gamma_1^w\varphi + \mathbf{C}\varphi].$$

The boundedness of $\theta_\eta^\varepsilon(x)\varphi(w)$ follows from the form of the operators $\hat{\Gamma}_1$ and R_0.

The proof of the theorem is completed using Lemma 3.12 and Theorem 6.3 in Korolyuk and Limnios (2005).

4

Stochastic Approximation Procedure

4.1. Markov environment

4.1.1. *Jumping SAP in averaging scheme*

Jumping SAP in a Markov environment in scheme series can be given by the relation

$$u^\varepsilon(t) = u + \varepsilon \sum_{k=0}^{\nu(t/\varepsilon)-1} a_k^\varepsilon C(u_k^\varepsilon, x_k^\varepsilon), u^\varepsilon(0) = u \qquad [4.1]$$

(here, we assume $\sum_{k=0}^{-1} a_k^\varepsilon C(u_k^\varepsilon, x_k^\varepsilon) = 0$),

where the regression function $C(u,x) = (C_k(u,x), k = \overline{1,d})$, $u \in R^d, x \in X$, satisfies the conditions of the existing global solution for accompanying systems:

$$du^x(t)/dt = C(u^x(t), x), x \in X.$$

The external environment is described by a uniformly ergodic Markov process $x(t), t \geq 0$, on a standard phase space $(X, \boldsymbol{X})$ with the generator Q and its potential R_0.

In SAP [4.1], the normalizing sequence $a_n^\varepsilon, n \geq 0$, is determined by the control function values $a(t), t \geq 0$,

$$a_n^\varepsilon := a(\tau_n^\varepsilon), \tau_n^\varepsilon := \varepsilon \tau_n, n \geq 0,$$

where $\tau_n, n \geq 0$, are renewal moments of embedded Markov chain

$$x_n := x(\tau_n), n \geq 0,$$

with the counting process $\nu(t) := \max\{n : \tau_n \leq t\}, t \geq 0$.

In SAP [4.1], true embeddings hold

$$u_n^\varepsilon = u^\varepsilon(\tau_n^\varepsilon), x_n^\varepsilon = x(\tau_n^\varepsilon), \tau_n^\varepsilon = \varepsilon\tau_n, n \geq 0. \quad [4.2]$$

The convergence of SAP [4.1] is considered under the exponential stability conditions of the averaging system

$$\frac{du(t)}{dt} = C(u(t)), C(u) = q \int_X \rho(dx)C(u,x), \quad [4.3]$$

where $\rho(B), B \in \boldsymbol{X}$, is a stationary distribution of embedded Markov chain x_n, $n \geq 0$.

THEOREM 4.1.– *We assume that the Lyapunov function $V(u) \in C^3(R^d)$, $u \in R^d$, of the averaging system [4.3] has the following properties:*

$C1$: *it provides the exponential stability of the averaging system [4.3]*

$$C(u)V'(u) \leq -c_0 V(u), c_0 > 0;$$

$C2$: *for $\overline{C}(u) := \max_{x \in X} C(u,x)$, there hold the estimates*

$$\left|\overline{C}(u)V'(u)\right| \leq c_1(1 + V(u)), c_1 > 0,$$

$$\left|\overline{C}(u)[\overline{C}(u)V'(u)]'\right| \leq c_2(1 + V(u)), c_2 > 0,$$

$$\left|\overline{C}(u)\overline{C}(u)[\overline{C}(u)V'(u)]''\right| \leq c_3(1 + V(u)), c_3 > 0.$$

We also require that the normalizing function $a(t)$ be monotonically decreasing, positive and there hold the conditions:

$$C3 : \int_0^\infty a(t)dt = \infty, \int_0^\infty a^2(t)dt < \infty.$$

Then, for each positive $\varepsilon \leq \varepsilon_0$, ε_0 is small enough, SAP [4.1] convergence with probability 1 to a unique equilibrium point of the averaging evolutionary system [4.3]:

$$P\{\lim_{t \to \infty} u^\varepsilon(t) = 0\} = 1.$$

Furthermore, we consider a two-component Markov process

$$u^\varepsilon(t), x_t^\varepsilon := x(t/\varepsilon), t \geq 0. \quad [4.4]$$

LEMMA 4.1.– *The Markov process [4.4] is determined by the generator*

$$\boldsymbol{L}_t^\varepsilon \varphi(u,x) = \varepsilon^{-1} Q\varphi(u,x) + \varepsilon^{-1} q(x)\boldsymbol{C}_t^\varepsilon(x)\varphi(u,x), \quad [4.5]$$

where

$$\boldsymbol{C}_t^\varepsilon(x)\varphi(u,x) = \int_X P(x,dy)[\varphi(u+\varepsilon a(t)C(u,x),y) - \varphi(u,y)]. \quad [4.6]$$

PROOF.– The generator $\boldsymbol{L}_t^\varepsilon$ of the Markov process [4.4] is determined by the relation

$$\lim_{\Delta\to 0} \Delta^{-1}\{E[\varphi(u^\varepsilon(t+\Delta), x_{t+\Delta}^\varepsilon)|u^\varepsilon(t) = u, x_t^\varepsilon = x] - \varphi(u,x)\} = \boldsymbol{L}_t^\varepsilon \varphi(u,x). \quad [4.7]$$

We calculate the conditional expectation

$$\begin{aligned} &E[\varphi(u^\varepsilon(t+\Delta), x_{t+\Delta}^\varepsilon)|u^\varepsilon(t) = u, x_t^\varepsilon = x] \\ &= E_{u,x}[\varphi(u^\varepsilon(t+\Delta), x_{t+\Delta}^\varepsilon)] \\ &= E_{u,x}[\varphi(u^\varepsilon(t+\Delta), x_{t+\Delta}^\varepsilon)][I(\theta_x \le \varepsilon^{-1}\Delta) + I(\theta_x > \varepsilon^{-1}\Delta)] \\ &= E_{u,x}[\varphi(u^\varepsilon(t+\Delta), x_{t+\Delta}^\varepsilon)]I(\theta_x \le \varepsilon^{-1}\Delta) \\ &\quad + E_{u,x}[\varphi(u^\varepsilon(t+\Delta), x_{t+\Delta}^\varepsilon)]I(\theta_x > \varepsilon^{-1}\Delta). \end{aligned} \quad [4.8]$$

Since $I(\theta_x \le \varepsilon^{-1}\Delta) = 1 - e^{-\varepsilon^{-1}q(x)\Delta} = \varepsilon^{-1}q(x)\Delta + o(\Delta)$, for the first summand in [4.8], we have

$$\begin{aligned} &E_{u,x}[\varphi(u^\varepsilon(t+\Delta), x_{t+\Delta}^\varepsilon)]I(\theta_x \le \varepsilon^{-1}\Delta) \\ &= E_{u,x}[\varphi(u+\Delta u^\varepsilon(t), x_{t+\Delta}^\varepsilon)]\varepsilon^{-1}q(x)\Delta + o(\Delta), \end{aligned} \quad [4.9]$$

where evolutionary increments $\Delta u^\varepsilon(t) = u^\varepsilon(t+\Delta) - u^\varepsilon(t)$ are determined by procedure [4.1] in terms of expectation [4.8] as the relation

$$\Delta u^\varepsilon(t) = \varepsilon a(t)C(u,x).$$

While $I(\theta_x > \varepsilon^{-1}\Delta) = e^{-\varepsilon^{-1}q(x)\Delta} = 1 - \varepsilon^{-1}q(x)\Delta + o(\Delta)$, for the second summand in [4.8], we obtain

$$\begin{aligned} &E_{u,x}[\varphi(u^\varepsilon(t+\Delta), x_{t+\Delta}^\varepsilon)]I(\theta_x > \varepsilon^{-1}\Delta) \\ &= E_{u,x}[\varphi(u+\Delta u^\varepsilon(t), x_{t+\Delta}^\varepsilon)][1 - \varepsilon^{-1}q(x)\Delta + o(\Delta)]. \end{aligned}$$

Since at $\Delta \to 0$ for the last expression $\Delta u^\varepsilon(t) \to 0$, $x^\varepsilon_{t+\Delta} \to x$, we have

$$E_{u,x}[\varphi(u^\varepsilon(t+\Delta), x^\varepsilon_{t+\Delta})]I(\theta_x > \varepsilon^{-1}\Delta)$$
$$= \varphi(u,x)[1 - \varepsilon^{-1}q(x)\Delta] + o(\Delta). \quad [4.10]$$

Therefore, considering [4.9] and [4.10], from [4.8], we have

$$E_{u,x}[\varphi(u^\varepsilon(t+\Delta), x^\varepsilon_{t+\Delta})]$$
$$= \varphi(u,x) - \varepsilon^{-1}q(x)\varphi(u,x)\Delta$$
$$+\varepsilon^{-1}q(x)E_{u,x}\varphi(u + \varepsilon a(t)C(u,x), y)\Delta + o(\Delta), \quad [4.11]$$

with legitimate replacement $x^\varepsilon_{t+\Delta}$ to $y \in X$.

We use the result [4.11] in [4.7] to find the generator $\boldsymbol{L}^\varepsilon_t$ in a sequence of transformations:

$$\begin{aligned}\boldsymbol{L}^\varepsilon_t\varphi(u,x) &= \lim_{\Delta\to 0} \Delta^{-1}[\varphi(u,x) - \varepsilon^{-1}q(x)\varphi(u,x)\Delta \\ &\quad + \varepsilon^{-1}q(x)E_{u,x}\varphi(u+\varepsilon a(t)C(u,x), y)\Delta - \varphi(u,x) + o(\Delta) \\ &= \varepsilon^{-1}q(x)\int_X P(x,dy)[\varphi(u+\varepsilon a(t)C(u,x),y) - \varphi(u,x)] \\ &= \varepsilon^{-1}q(x)\int_X P(x,dy)[\varphi(u,y) - \varphi(u,x)] \\ &\quad + \varepsilon^{-1}q(x)\int_X P(x,dy)[\varphi(u+\varepsilon a(t)C(u,x),y) - \varphi(u,y)],\end{aligned}$$

i.e. we obtain [4.5] and the lemma is proved.

COROLLARY 4.1.– *The generator $\boldsymbol{L}^\varepsilon_t$ on the test functions $\varphi(u,\cdot) \in C^2(R^d)$ has the asymptotic representation*

$$\boldsymbol{L}^\varepsilon_t\varphi(u,x) = \varepsilon^{-1}Q\varphi(u,x) + a(t)Q_1(x)\varphi(u,x)$$
$$+ \varepsilon a^2(t)\theta^\varepsilon_1(x)\varphi(u,x), \quad [4.12]$$

where

$$Q_1(x)\varphi(u,x) = C(u,x)Q_0\varphi'_u(u,x), \quad [4.13]$$

$$\theta^\varepsilon_1(x)\varphi(u,x) = \frac{1}{2}C^2(u,x)Q_0\varphi''_u(\theta u, x), 0 \le \theta \le 1, \quad [4.14]$$

and

$$Q_0\varphi(x) := q(x)\boldsymbol{P}\varphi(x),$$

$$\boldsymbol{P}\varphi(x) := \int_X P(x,dy)\varphi(y).$$

PROOF.– We use the smoothness of the function $\varphi(u,x)$ in the calculation of the operator $\boldsymbol{C}_t^\varepsilon(x)$ in [4.6]:

$$\boldsymbol{C}_t^\varepsilon\varphi(u,x) = \int_X P(x,dy)[\varphi(u+\varepsilon a(t)C(u,x),y) - \varphi(u,y)]$$

$$= \int_X P(x,dy)[\varepsilon a(t)C(u,x)\varphi_u'(u,y) + \varepsilon^2 a^2(t)\frac{1}{2}C^2(u,x)\varphi_u''(\theta u,y)], 0 \le \theta \le 1.$$

Therefore, we have the representation [4.12].

In [4.12], $\theta_1^\varepsilon(x)\varphi(u,x)$ is the remainder term and

$$\|\theta_1^\varepsilon(x)\varphi(u,x)\| \le C, C > 0,$$

that is,

$$\varepsilon a^2(t)\|\theta_1^\varepsilon(x)\varphi(u,x)\| \to 0, \varepsilon \to 0. \qquad [4.15]$$

It follows from [4.14] and the smoothness of the function $\varphi(u,x)$.

We now solve the singular perturbation problem for the operator $\boldsymbol{L}_t^\varepsilon$ on the perturbed Lyapunov function

$$V^\varepsilon(u,x) = V(u) + \varepsilon a(t)V_1(u,x), \qquad [4.16]$$

where $V(u)$ is the Lyapunov function for the averaging system [4.3].

LEMMA 4.2.– *The generator $\boldsymbol{L}_t^\varepsilon$ on the perturbed Lyapunov function $V^\varepsilon(u,x)$ with $V(u) \in C^3(R^d)$ has the asymptotic representation*

$$\boldsymbol{L}_t^\varepsilon V^\varepsilon(u,x) = a(t)C(u)V'(u) + \varepsilon a^2(t)\theta_L^\varepsilon(x)V(u), \qquad [4.17]$$

where

$$\theta_L^\varepsilon(x)V(u) = \theta_0^\varepsilon(x)V(u) + \theta_1^\varepsilon(x)V(u), \qquad [4.18]$$

$$\theta_0^\varepsilon(x)V(u) = C(u,x)[[q(x)R_0 - I]\widetilde{C}_u'(u,x)V'(u) + \Pi\widetilde{C}_u'(u,x)V'(u)$$
$$+ [q(x)R_0 - I]\widetilde{C}(u,x)V''(u)]\varphi_u''(\theta u,y)], 0 \le \theta \le 1, \qquad [4.19]$$

$$\theta_0^\varepsilon(x)V(u) = \frac{1}{2}C^2(u,x)q(x)V''(\theta u), 0 \le \theta \le 1. \qquad [4.20]$$

PROOF.– First, we solve the singular perturbation problem for the truncated operator

$$\boldsymbol{L}_{t0}^{\varepsilon} = \varepsilon^{-1}Q + a(t)Q_1(x).$$

We represent the decomposition of the operator $\boldsymbol{L}_{t0}^{\varepsilon}$ on the functions $V^{\varepsilon}(u,x)$ in the form

$$\begin{aligned}\boldsymbol{L}_{t0}^{\varepsilon}V^{\varepsilon}(u,x) &= [\varepsilon^{-1}Q + \varepsilon a(t)Q_1(x)][V(u) + a(t)V_1(u,x)]\\ &= \varepsilon^{-1}QV(u)a(t)[QV_1(u,x) + Q_1(x)V(u)] + \varepsilon a^2(t)Q_1(x)V_1(u,x).\end{aligned}$$

From the solvability condition of the last decomposition, we have

$$QV_1(u,x) + Q_1(x)V(u) = \widehat{Q}_1V(u), \qquad [4.21]$$

where

$$\widehat{Q}_1\Pi = \Pi Q_1(x)\Pi. \qquad [4.22]$$

Then, we calculate the right-hand side of [4.22]. From the representation [4.13], we have

$$\begin{aligned}\Pi Q_1(x)\Pi V(u) &= \Pi C(u,x)Q_0\Pi V'(u)\\ &= \Pi C(u,x)q(x)\boldsymbol{P}\Pi V'(u)\\ &= \Pi C(u,x)q(x)V'(u)\\ &= \int_X \pi(dx)q(x)C(u,x)V'(u)\\ &= q\int_X \rho(dx)C(u,x)V'(u)\\ &= C(u)V'(u).\end{aligned}$$

Therefore, from [4.21], we obtain

$$QV_1(u,x) = (\widehat{Q}_1 - Q_1(x))V(u).$$

According to the property of the potential R_0 from the last expression, we have the following solution:

$$V_1(u,x) = R_0\widetilde{Q}_1(x)V(u), \qquad [4.23]$$

where

$$\widetilde{Q}_1(x) = Q_1(x) - \widehat{Q}_1. \qquad [4.24]$$

Furthermore, we calculate the right-hand side of [4.24] according to [4.13] and [4.22]

$$[Q_1(x) - \widehat{Q}_1]V(u) = [C(u,x)Q_0 - C(u)]V'(u)$$
$$= [C(u,x)q(x)\boldsymbol{P} - C(u)]V'(u) = [C(u,x)q(x) - C(u)]V'(u).$$

So, for perturbation $V_1(u,x)$ of the Lyapunov function $V(u)$ from [4.23], we obtain the representation

$$V_1(u,x) = R_0\widetilde{\boldsymbol{C}}(x)V(u), \qquad [4.25]$$

where

$$\widetilde{\boldsymbol{C}}(x)V(u) = \widetilde{C}(u,x)V'(u),$$
$$\widetilde{C}(u,x) = C(u,x)q(x) - C(u).$$

We now use [4.25] for calculating the multiplier

$$\theta_0^\varepsilon(x)V(u) := Q_1(x)V_1(u,x)$$

of decomposition for an operator $\boldsymbol{L}_{t0}^\varepsilon\, V^\varepsilon(u,x)$:

$$\theta_0^\varepsilon(x)V(u) = C(u,x)Q_0[R_0\widetilde{\boldsymbol{C}}(x)V(u)]' =$$
$$= C(u,x)q(x)\boldsymbol{P}R_0[\widetilde{C}(u,x)V'(u)]' =$$
$$= C(u,x)q(x)\boldsymbol{P}R_0[\widetilde{C}'_u(u,x)V'(u) + \widetilde{C}(u,x)V''(u)]. \qquad [4.26]$$

Next, we continue the calculation in [4.26], taking into account $q(x)\boldsymbol{P}R_0 = q(x)R_0 + \Pi - I$:

$$\begin{aligned}\theta_0^\varepsilon(x)V(u) &= C(u,x)[q(x)R_0 + \Pi - I][\widetilde{C}'_u(u,x)V'(u) + \widetilde{C}(u,x)V''(u)]\\ &= C(u,x)[q(x)R_0\widetilde{C}'_u(u,x)V'(u) + q(x)R_0\widetilde{C}(u,x)V''(u)]\\ &\quad + C(u,x)[\Pi\widetilde{C}'_u(u,x)V'(u) + \Pi\widetilde{C}(u,x)V''(u)]\\ &\quad - C(u,x)[\widetilde{C}'_u(u,x)V'(u) + \widetilde{C}(u,x)V''(u)]\\ &= C(u,x)[[q(x)R_0 - I]\widetilde{C}'_u(u,x)V'(u) + \Pi\widetilde{C}'_u(u,x)V'(u)\\ &\quad + [q(x)R_0 - I]\widetilde{C}(u,x)V''(u).\end{aligned}$$

Therefore, we have the representation

$$\boldsymbol{L}_{t0}^\varepsilon V^\varepsilon(u,x) = a(t)C(u)V'(u) + \varepsilon a^2(t)\theta_0^\varepsilon(x)V(u), \qquad [4.27]$$

where $\theta_0^\varepsilon(x)V(u)$ has the form [4.19].

We note that from the smoothness of the functions $C(u,x), V(u)$, and the limitness of the operator R_0 according to [4.19], we obtain

$$\|\theta_0^\varepsilon(x)V(u)\| < C, C > 0, x \in X. \qquad [4.28]$$

Returning to the full representation of the generator $\boldsymbol{L}_t^\varepsilon$, we consider it in the form

$$\boldsymbol{L}_t^\varepsilon = \boldsymbol{L}_{t0}^\varepsilon + \varepsilon a^2(t)\theta_1^\varepsilon(x).$$

Therefore, we have decomposition on functions [4.16]

$$\begin{aligned}\boldsymbol{L}_t^\varepsilon V^\varepsilon(u,x) &= \boldsymbol{L}_{t0}^\varepsilon[V(u) + \varepsilon a(t)V_1(u,x)]\\ &\quad + \varepsilon a^2(t)\theta_1^\varepsilon(x)[V(u) + \varepsilon a(t)V_1(u,x)],\end{aligned} \qquad [4.29]$$

where $V_1(u,x)$ is calculated by formula [4.25].

Substituting [4.27] into [4.29], we have

$$\begin{aligned}\boldsymbol{L}_t^\varepsilon V^\varepsilon(u,x) &= a(t)C(u)V'(u) + \varepsilon a^2(t)\theta_0^\varepsilon(x)V(u)\\ &\quad + \varepsilon a^2(t)\theta_1^\varepsilon(x)V(u) + \varepsilon^2 a^3(t)\theta_1^\varepsilon(x)R_0\widetilde{C}(u,x)V'(u).\end{aligned} \qquad [4.30]$$

Let us calculate the last two summand in [4.30]:

$$\theta_1^\varepsilon(x)V(u) = \frac{1}{2}C^2(u,x)q(x)V''(\theta u), 0 \le \theta \le 1.$$

$$\theta_1^\varepsilon(x)R_0\widetilde{C}(u,x)V'(u) = \frac{1}{2}C^2(u,x)Q_0R_0[\widetilde{C}(u,x)V'(u)]''_u|_{u=\theta u}.$$

Since the functions $C(u,x), V(u)$ are smooth and the operators R_0 and Q_0 are limited, we obtain

$$\|\theta_1^\varepsilon(x)R_0\widetilde{C}(u,x)V'(u)\| \le M, M > 0, x \in X,$$

and it gives opportunity to ignore the corresponding summand in decomposition [4.30] according to the second-order smallness of the variable ε.

As a result, from [4.30], we obtain [4.17] and the lemma is proved.

PROOF OF THEOREM.– We provide it in several steps. First, we establish the key inequality

LEMMA 4.3.– *The generator $\boldsymbol{L}_t^\varepsilon$ on the perturbed Lyapunov function [4.16] under theorem conditions allows for the evaluation*

$$\boldsymbol{L}_t^\varepsilon V^\varepsilon(u,x) \le -\delta a(t)V(u), \delta > 0. \qquad [4.31]$$

PROOF.– Using evaluations $C2$ for the remainder term $\theta_L^\varepsilon(x)V(u)$, condition 1 of exponential stability for the system [4.3] as well as the monotonicity and the boundedness of the normalizing function $a(t)$, we have

$$\begin{aligned} \boldsymbol{L}_t^\varepsilon V^\varepsilon(u,x) &= a(t)C(u)V'(u) + \varepsilon a^2(t)\theta_L^\varepsilon(x)V(u) \\ &\leq -a(t)(\delta^* - \varepsilon c)V(u) \leq -a(t)\delta V(u), \delta > 0. \end{aligned} \quad [4.32]$$

The last inequality holds true at each $\varepsilon \leq \varepsilon_0$, where ε_0 satisfies the inequality $\varepsilon_0 < \delta^*/c$.

Second, an explicit form of perturbation [4.24], as well as condition $C2$ of the theorem, gives an estimate of the Lyapunov perturbed function

$$0 < (1 - \varepsilon a(t)c)V(u) \leq V^\varepsilon(u,x) \leq (1 + \varepsilon a(t)c)V(u).$$

Third, we consider that the process

$$V_n^\varepsilon := V^\varepsilon(u_n^\varepsilon, x_n^\varepsilon), n \geq 0, \quad [4.33]$$

is a non-negative martingale. For this, we note that the extended renewal Markov process [4.7] generates on the test functions $\varphi(u,x), u \in R^d, x \in X$, martingale

$$\mu_{n+1} = \varphi(u_{n+1}^\varepsilon, x_{n+1}^\varepsilon) - \varphi(u_0, x_0) - \varepsilon \sum_{k=0}^{n} \theta_{k+1} \boldsymbol{L}_{\tau_k^\varepsilon}^\varepsilon \varphi(u_k^\varepsilon, x_k^\varepsilon), \quad [4.34]$$

relative to the flow σ-algebras $F_n^\varepsilon = \sigma\{u_k^\varepsilon, x_k^\varepsilon, \tau_k^\varepsilon, 0 \leq k \leq n\}, n \geq 0$. Indeed, we have

$$E[\mu_{n+1} - \mu_n \,|F_n^\varepsilon] = E_{F_n^\varepsilon}[\varphi_{n+1}^\varepsilon - \varphi_n^\varepsilon] - E_{F_n^\varepsilon}[\varepsilon \theta_{n+1} \boldsymbol{L}_{\tau_n^\varepsilon}^\varepsilon \varphi_n^\varepsilon]. \quad [4.35]$$

in the notation $\varphi_n^\varepsilon := \varphi(u_n^\varepsilon, x_n^\varepsilon)$

In accordance with the definition of the compensating operator [4.8], we calculate the second summand in [4.35]:

$$\begin{aligned} & E_{F_n^\varepsilon}[\varepsilon \theta_{n+1} \boldsymbol{L}_{\tau_n^\varepsilon}^\varepsilon \varphi_n^\varepsilon] = \varepsilon g(x_n^\varepsilon) \boldsymbol{L}_{\tau_n^\varepsilon}^\varepsilon \varphi_n^\varepsilon = \varepsilon g(x_n^\varepsilon)\varepsilon^{-1} q(x_n^\varepsilon) E_{F_n^\varepsilon}[\varphi_{n+1}^\varepsilon - \varphi_n^\varepsilon] \\ & = E_{F_n^\varepsilon}[\varphi_{n+1}^\varepsilon - \varphi_n^\varepsilon]. \end{aligned}$$

Therefore, according to [4.35], we have martingale property in [4.34]:

$$E[\mu_{n+1} - \mu_n \,|F_n^\varepsilon] = 0, n \geq 0. \quad [4.36]$$

We use [4.36] to prove the next lemma.

Furthermore, in [4.34], we consider the perturbed Lyapunov function $V^\varepsilon(u,x)$ instead of any test function $\varphi(u,x), u \in R^d, x \in X$.

LEMMA 4.4.– *The process [4.33] is a non-negative supermartingale.*

PROOF.– We use martingale characterization of process [4.33], namely, justify the truth of the presentation

$$E[V_{n+1}^{\varepsilon} \,|F_n^{\varepsilon}] = V_0^{\varepsilon} + \varepsilon E[\sum_{k=0}^{n} \theta_{k+1}\, \boldsymbol{L}_{\tau_k^{\varepsilon}}^{\varepsilon} V_k^{\varepsilon}] + \mu_n^{\varepsilon}, \qquad [4.37]$$

since from [4.36], we have $E[\mu_{n+1}^{\varepsilon} \,|F_n^{\varepsilon}] = \mu_n^{\varepsilon}$. Substituting into [4.37] instead of μ_n^{ε} its value from [4.34], we obtain

$$E[V_{n+1}^{\varepsilon} \,|F_n^{\varepsilon}] = V_0^{\varepsilon} + \varepsilon g(x_n^{\varepsilon})\boldsymbol{L}_{\tau_n^{\varepsilon}}^{\varepsilon} V_n^{\varepsilon} + \varepsilon[\sum_{k=0}^{n-1} \theta_{k+1}\boldsymbol{L}_{\tau_k^{\varepsilon}}^{\varepsilon} V_k^{\varepsilon}]$$

$$+ V_n^{\varepsilon} - V_0^{\varepsilon} - \varepsilon[\sum_{k=0}^{n-1} \theta_{k+1}\boldsymbol{L}_{\tau_k^{\varepsilon}}^{\varepsilon} V_k^{\varepsilon}].$$

There we have $E[V_{n+1}^{\varepsilon} \,|F_n^{\varepsilon}] = V_n^{\varepsilon} + \varepsilon g(x_n^{\varepsilon})\boldsymbol{L}_{\tau_n^{\varepsilon}}^{\varepsilon} V_n^{\varepsilon}$, and with key inequality [4.31], we obtain $E[V_{n+1}^{\varepsilon} \,|F_n^{\varepsilon}] \geq V_n^{\varepsilon}, n \geq 0$, i.e. lemma is proved.

Using estimates [4.31] and [4.32] as well as Lemma 4.4 completes proof of Theorem 4.1.

4.1.2. *Jumping SAP under diffusion approximating scheme*

We consider jumping SAP, determined by the relation (we assume that $\sum_{k=0}^{-1} a_k^{\varepsilon} C^{\varepsilon}(u_k^{\varepsilon}, x_k^{\varepsilon}) := 0$):

$$u^{\varepsilon}(t) = u + \varepsilon \sum_{n=1}^{\nu(t/\varepsilon^2)-1} a_n^{\varepsilon} C^{\varepsilon}(u_n^{\varepsilon}, x_n^{\varepsilon}), u^{\varepsilon}(0) = u, \qquad [4.38]$$

where $\nu(t) := \max\{n, \tau \leq t\}$ is the counting process of jumps $\tau_n, n \geq 1$, of the Markov process $x(t), t \geq 0$.

Hence, $x_n = x(\tau_n), n \geq 1$ is an embedded Markov chain and $u_n^{\varepsilon} = u^{\varepsilon}(\tau_n)$, $n \geq 1$, is a discrete SAP that we consider in the Euclidean space R^d in the form

$$u_{n+1}^{\varepsilon} - u_n^{\varepsilon} = \varepsilon a_n^{\varepsilon} C^{\varepsilon}(u_n^{\varepsilon}, x_n^{\varepsilon}), n \geq 0,$$

where the function $a(t)$ generates the sequence $a_n = a(\tau_n), n \geq 1$, and the sequence of points $u_n^{\varepsilon} \in R^d$ is that

$$\lim_{n \to \infty} u_n^{\varepsilon} = u_0$$

almost everywhere, when $\varepsilon > 0$ is small enough.

The regression function

$$C^{\varepsilon}(u,x) = C_0(u,x) + \varepsilon^{-1}C(u,x) \quad [4.39]$$

depends on the external environment states $x \in X$, whose change is given by ergodic Markov chain x_n,$n \geq 0$, with transition probabilities

$$P(x,B) = P\{x_{n+1} \in B/x_n = x\}, x \in X, B \in \boldsymbol{X}, \quad [4.40]$$

$\varepsilon > 0$ is a small series parameter and $(X, \boldsymbol{X})$ is a phase space of states of the Markov process $x(t), t \geq 0$.

Furthermore, we study the convergence of SAP [4.41] with the perturbed regression function [4.39] under the balance conditions

$$\int_X \rho(dx)C_0(u,x) = 0. \quad [4.41]$$

Averaging SAP under the balance conditions [4.41] is determined by the stochastic differential equation

$$du(t) = a(t)C(u(t))dt + a^2(t)d\zeta(t), \quad [4.42]$$

where

$$d\zeta(t) = b(u(t))dt + \frac{1}{2}SpB((u(t))dw_t \quad [4.43]$$

is a diffusion process with shift $b(u)$ and covariation matrix $B(u)$, w_t is a Wiener process and the averaging regression function $C(u)$ is determined by the formula

$$C(u) = q\int_X \rho(dx)C(u,x). \quad [4.44]$$

The coefficients of the perturbed diffusion process $\zeta(t), t \geq 0$, the shift $b(u)$ and the covariation matrix $B(u)$, are determined by the relations

$$b(u) = q\int_X \rho(dx)b(u,x), \quad [4.45]$$

$$B(u) = q\int_X \rho(dx)B(u,x), \quad [4.46]$$

where

$$b(u,x) := C_0(u,x)R_0q(x)C_0'(u,x), \quad [4.47]$$

$$B(u,x) := 2C_0(u,x)R_0q(x)C_0(u,x) - C_0^2(u,x). \quad [4.48]$$

We input majorants for the sum of the regression function $C(u, x)$ and its perturbation $C_0(u, x)$ to formulate the main result of this sub-section

$$\overline{C}(u) := \max_{x \in X} (|C(u, x)| + |C_0(u, x)|) \quad [4.49]$$

and additional functions

$$w_k(u) := \overline{C}(u) w'_{k-1}(u), \quad k = 1, 2, 3, 4, 5; \quad [4.50]$$

$$w_0(u) := \overline{C}(u) V'(u), \quad [4.51]$$

where $V(u)$ is the Lyapunov function for the averaging dynamical system

$$du(t)/dt = C(u(t)). \quad [4.52]$$

By convergence of SAP [4.41], we formulate under the conditions of the uniform exponential stability of the solution $u(t)$ for the system [4.52], i.e. for [4.52], there exists a unique equilibrium point u_0, $C(u_0) = 0$, and trajectories of the averaging system [4.52] converge to this point.

THEOREM 4.2.– *We assume that the existence of the Lyapunov function $V(u)$ for the averaging dynamical system [4.52] has limited derivatives for the fifth-order inclusive and it satisfies the following conditions:*

$$C1 : C(u)V'(u) \leq -cV(u);$$

$$C2 : w_k(u) \leq c_k(1 + V(u)), \quad k = 1, 2, 3, 4, 5.$$

In addition, the regression functions $C(u, x)$ and $C_0(u, x)$ of SAP [4.41] also have four derivatives at u that are uniformly bounded at $x \in X$, and the function $C_0(u, x)$ satisfies the balance condition [4.52].

The normalization function $a(t) > 0$ is selected so that the following conditions are fulfilled:

$$\int_0^\infty a(t)dt = \infty, \quad \int_0^\infty a^2(t)dt < \infty.$$

Then, jumping SAP [4.41] as well as averaging SAP [4.42], [4.43] converge to an equilibrium point of the dynamical system [4.52] with probability 1.

Proof of theorem is based on using a singular perturbation problem solution for the reducible-invertible operator Q that determines the ergodic Markov process.

LEMMA 4.5.– *The generator of the two-component Markov process $u^\varepsilon(t)$, $x^\varepsilon(t) = x(t/\varepsilon^2), t \geq 0$ can be represented in the form*

$$\mathbf{L}_t^\varepsilon = \varepsilon^{-2}[Q + \mathbf{C}_t^\varepsilon(x)]. \quad [4.53]$$

Here

$$\boldsymbol{C}_t^\varepsilon(x)\varphi(u,x) = q(x)\int\limits_X P(x,dy)[\varphi(u+\varepsilon a(t)C^\varepsilon(u,x),y) - \varphi(u,y)]. \quad [4.54]$$

PROOF.– We use the definition of the generator of the Markov process to construct the generator $\boldsymbol{L}_t^\varepsilon$.

$$\boldsymbol{L}_t^\varepsilon\varphi(u,x) := \lim_{\Delta\to 0}\Delta^{-1}E_{u,x}[\varphi(u^\varepsilon(t+\Delta),x^\varepsilon(t+\Delta)) - \varphi(u,x)]. \quad [4.55]$$

We consider the conditional expectation $E_{u,x}$, using [4.38], as well as $x^\varepsilon(t) = x(t/\varepsilon^2)$:

$$\begin{aligned}
&E_{u,x}\varphi(u^\varepsilon(t+\Delta),x^\varepsilon(t+\Delta))\\
&= E_{u,x}[\varphi(u+\varepsilon a(t)C^\varepsilon(u,x),x^\varepsilon(t+\Delta))I(\theta_x \le \varepsilon^{-2}\Delta)\\
&\quad + \varphi(u,x^\varepsilon(t))I(\theta_x > \varepsilon^{-2}\Delta)],
\end{aligned}$$

where θ_x is a staying time in state x.

Since

$$I(\theta_x \le \varepsilon^{-2}\Delta) = 1 - e^{-\varepsilon^{-2}\Delta q(x)}, \quad I(\theta_x > \varepsilon^{-2}\Delta) = e^{-\varepsilon^{-2}\Delta q(x)},$$

where $q(x)$ is the intensity of the staying time in state x, then

$$\begin{aligned}
&E_{u,x}\varphi(u^\varepsilon(t+\Delta),x^\varepsilon(t+\Delta))\\
&= E_{u,x}[\varphi(u+\varepsilon a(t)C^\varepsilon(u,x),x^\varepsilon(t+\Delta))(1-e^{-\varepsilon^{-2}\Delta q(x)})\\
&\quad + \varphi(u,x^\varepsilon(t))e^{-\varepsilon^{-2}\Delta q(x)}]\\
&= E_{u,x}[\varphi(u+\varepsilon a(t)C^\varepsilon(u,x),x^\varepsilon(t+\Delta))\varepsilon^{-2}\Delta q(x) + o_1(\Delta^2)\\
&\quad + \varphi(u,x) - \varphi(u,x)\varepsilon^{-2}\Delta q(x) + o_2(\Delta^2)]\\
&= \varepsilon^{-2}\Delta q(x)E_{u,x}[\varphi(u+\varepsilon a(t)C^\varepsilon(u,x),x^\varepsilon(t+\Delta)) - \varphi(u,x)] + o(\Delta^2).
\end{aligned}$$

Now we consider the first summand in the last expression. Since

$$\begin{aligned}
&\varepsilon^{-2}\Delta q(x)E_{u,x}[\varphi(u+\varepsilon a(t)C^\varepsilon(u,x),x^\varepsilon(t+\Delta)) - \varphi(u,x)]\\
&= \varepsilon^{-2}\Delta q(x)\int\limits_X P(x,dy)[\varphi(u+\varepsilon a(t)C^\varepsilon(u,x),y) - \varphi(u,x)],
\end{aligned}$$

then, according to [4.55], for the generator $\boldsymbol{L}_t^\varepsilon$, we have

$$\boldsymbol{L}_t^\varepsilon\varphi(u,x) = q(x)\varepsilon^{-2}\int\limits_X P(x,dy)[\varphi(u+\varepsilon a(t)C^\varepsilon(u,x),y) - \varphi(u,x)].$$

We make some transformations to highlight the Markov process generator Q:

$$\begin{aligned}
\boldsymbol{L}_t^\varepsilon\varphi(u,x) &= \varepsilon^{-2}q(x)\int\limits_X [P(x,dy)\varphi(u+\varepsilon a(t)C^\varepsilon(u,x),y) - P(x,dy)\varphi(u,y) \\
&\quad + P(x,dy)\varphi(u,y) - Q(x,dy)\varphi(u,x)] \\
&= \varepsilon^{-2}q(x)\int\limits_X [P(x,dy)[\varphi(u,y) - \varphi(u,x)] \\
&\quad + \varepsilon^{-2}q(x)\int\limits_X P(x,dy)[\varphi(u+\varepsilon a(t)C^\varepsilon(u,x),y) - \varphi(u,y)],
\end{aligned}$$

i.e. we have [4.55] in the operator form.

LEMMA 4.6.– *The generator $\boldsymbol{L}_t^\varepsilon$ at three times continuously derivative at u functions $\varphi(u,x)$ has the form*

$$\begin{aligned}
\boldsymbol{L}_t^\varepsilon\varphi(u,x) &= \varepsilon^{-2}Q\varphi(u,x) + \varepsilon^{-1}a(t)Q_1(x)\varphi(u,x) \\
&\quad + a(t)Q_{21}(x)\varphi(u,x) + \frac{1}{2}a^2(t)Q_{22}(x)\varphi(u,x) \\
&\quad + \varepsilon a^2(t)\boldsymbol{G}_t^\varepsilon(x)\varphi(u,x), \qquad [4.56]
\end{aligned}$$

where

$$Q_1(x)\varphi(u,x) = C_0(u,x)Q_0\varphi'(u,x), \qquad [4.57]$$

$$\begin{aligned}
Q_{21}(x)\varphi(u,x) &= C(u,x)Q_0\varphi'(u,x), && [4.58] \\
Q_{22}(x)\varphi(u,x) &= C_0^2(u,x)Q_0\varphi''(u,x), && [4.59] \\
\boldsymbol{G}_t^\varepsilon(x)\varphi(u,x) &= G_1(x)\varphi(u,x) + a(t)G_2(x)\varphi(u,x) + \varepsilon G_3(x)\varphi(u,x), && [4.60] \\
G_1(x)\varphi(u,x) &= C(u,x)C_0(u,x)Q_0\varphi''(u,x), && [4.61] \\
G_2(x)\varphi(u,x) &= \frac{1}{6}C_0^3(u,x)Q_0\varphi'''(u,x), && [4.62] \\
G_3(x)\varphi(u,x) &= \frac{1}{2}C^2(u,x)Q_0\varphi'''(u,x), && [4.63] \\
Q_0 &= q(x)\boldsymbol{P}.
\end{aligned}$$

PROOF.– We decompose the function under the sign of the integral in [4.54] into the Taylor series at u:

$$\begin{aligned}
&\varphi(u+\varepsilon a(t)C^{\varepsilon}(u,x),y)-\varphi(u,y)\\
&=\varphi(u+\varepsilon a(t)C_0(u,x)+\varepsilon^2 a(t)C(u,x),y)-\varphi(u,y)\\
&=\varphi(u+\varepsilon a(t)C_0(u,x),y)+\varepsilon^2 a(t)C(u,x)\varphi'(u+\varepsilon a(t)C_0(u,x),y)\\
&\quad+\frac{1}{2}\varepsilon^4 a^2(t)C^2(u,x)\varphi''(\tilde{u},y)-\varphi(u,y).
\end{aligned} \qquad [4.64]$$

In [4.64] and hereinafter, $\tilde{u}$ is the corresponding midpoint of the remainder terms of decomposition into Taylor series.

Since

$$\varphi(u+\varepsilon a(t)C_0(u,x),y)=\varphi(u,y)+\varepsilon a(t)C_0(u,x)\varphi'(u,y)$$
$$+\frac{1}{2}\varepsilon^2 a^2(t)C_0^2(u,x)\varphi''(u,y)+\frac{1}{6}\varepsilon^3 a^3(t)C_0^3(u,x)\varphi_u'''(\tilde{u},y)$$

and

$$\varphi'(u+\varepsilon a(t)C_0(u,x),y)=\varphi'(u,y)+\varepsilon a(t)C_0(u,x)\varphi''(\tilde{u},y),$$

then in accordance with [4.64], we have

$$\varphi(u+\varepsilon a(t)C^{\varepsilon}(u,x),y)-\varphi(u,y)=\varepsilon a(t)C_0(u,x)\varphi'(u,y)$$
$$+\frac{1}{2}\varepsilon^2 a^2(t)C_0^2(u,x)\varphi''(u,y)+\varepsilon^2 a(t)C(u,x)\varphi'(u,y)$$
$$+\varepsilon^3 a^2(t)C(u,x)C_0(u,x)\varphi''(\tilde{u},y)+\frac{1}{6}\varepsilon^3 a^3(t)C_0^3(u,x)\varphi'''(\tilde{u},y)$$
$$+\frac{1}{2}\varepsilon^4 a^2(t)C^2(u,x)\varphi''(\tilde{u},y).$$

Consequently, the operator $\boldsymbol{C}_t^{\varepsilon}(x)$, according to [4.54], at the functions $\varphi(u,x)$, has the form

$$\begin{aligned}
\boldsymbol{C}_t^{\varepsilon}(x)\varphi(u,x)&=\varepsilon a(t)q(x)C_0(u,x)\int_X P(x,dy)\varphi'(u,y)\\
&\quad+\varepsilon^2[a(t)q(x)C(u,x)\int_X P(x,dy)\varphi'(u,y)\\
&\quad+\frac{1}{2}a^2(t)q(x)C_0^2(u,x)\int_X P(x,dy)\varphi''(u,y)]\\
&\quad+\varepsilon^3[a^2(t)C(u,x)C_0(u,x)q(x)\int_X P(x,dy)\varphi''(u,y)
\end{aligned}$$

$$+ \frac{1}{6} a^3(t) C_0^3(u,x) q(x) \int_X P(x,dy) \varphi'''(u,y)$$

$$+ \frac{1}{2} \varepsilon a^2(t) C^2(u,x) q(x) \int_X P(x,dy) \varphi''(u,y)].$$

According to the notations [4.57]–[4.63], we have [4.56].

When the function $\varphi(u,x)$ depends on only the variable u, from [4.57]–[4.63], we have:

$$Q_1(x)\varphi(u) = q(x) C_0(u,x) \varphi'(u), \quad [4.65]$$

$$Q_{21}(x)\varphi(u) = q(x) C(u,x) \varphi'(u), \quad [4.66]$$

$$Q_{22}(x)\varphi(u) = q(x) C_0^2(u,x) \varphi''(u), \quad [4.67]$$

$$\mathbf{G}_t^\varepsilon(x)\varphi(u) = q(x)[C(u,x) C_0(u,x) \varphi''(u) + \frac{1}{6} a(t) C_0^3(u,x) \varphi'''(u) + \frac{1}{2} \varepsilon^2 C^2(u,x) \varphi''(u)]. \quad [4.68]$$

Furthermore, we obtain the asymptotic representation of the operator $\boldsymbol{L}_t^\varepsilon$ on the perturbed Lyapunov test function

$$V^\varepsilon(u,x) = V(u) + \varepsilon a(t) V_1(u,x) + \varepsilon^2 a(t) V_2(u,x), \quad [4.69]$$

using the solution of the singular perturbation problem under the balance condition.

LEMMA 4.7.– *The generator [4.56] has the asymptotic representation*

$$\boldsymbol{L}_t^\varepsilon V^\varepsilon(u,x) = a(t) C(u) V'(u) + a^2(t) \boldsymbol{L}_0 V(u) + \varepsilon a^2(t) \boldsymbol{H}_t^\varepsilon(x) V(u), \quad [4.70]$$

on functions [4.69] with $V(u) \in C^5(R^d)$*, where*

$$\boldsymbol{L}_0 V(u) = b(u) V'(u) + \frac{1}{2} Sp[B(u) V''(u)], \quad [4.71]$$

and the remainder term $\boldsymbol{H}_t^\varepsilon(x) V(u)$ *is bounded.*

PROOF.– We prove the theorem in two steps.

Step 1. We consider the solution for the singular perturbation problem for the truncate operator [4.56], namely

$$\boldsymbol{L}_{t0}\varphi(u,x) = \varepsilon^{-2} Q\varphi(u,x) + \varepsilon^{-1} a(t) Q_1(x)\varphi(u,x) + a(x) Q_{21}(x)\varphi(u,x) + \frac{1}{2} a^2(t) Q_{22}(x)\varphi(u,x). \quad [4.72]$$

The representation of the operator $\boldsymbol{L}_{t0}$ on functions [4.69] has the form

$$\begin{aligned}\boldsymbol{L}_{t0}V^{\varepsilon}(u,x) &= \varepsilon^{-2}QV(u)\\ &+ \varepsilon^{-1}[QV_1(u,x) + a(t)Q_1(x)V(u)]\\ &+ QV_2(u,x) + a(t)Q_1(x)V_1(u,x) + a(t)Q_{21}(x)V(u)\\ &+ \frac{1}{2}a^2(t)Q_{22}(x)V(u) + \varepsilon\theta^{\varepsilon}_{t0}(x)V(u),\end{aligned} \qquad [4.73]$$

where

$$\begin{aligned}\theta^{\varepsilon}_{t0}(x)V(u) &= a(t)Q_1(x)V_2(u,x) + a(t)Q_{21}(x)V_1(u,x)\\ &+ \frac{1}{2}a^2(t)Q_{22}(x)V_1(u,x)\\ &+ \varepsilon a(t)Q_{21}(x)V_2(u,x) + \varepsilon\frac{1}{2}a^2(t)Q_{22}(x)V_2(u,x).\end{aligned} \qquad [4.74]$$

We define the conditions of solvability of decomposition [4.73]:

$$QV(u) = 0, \qquad [4.75]$$

$$QV_1(u,x) + a(t)Q_1(x)V(u) = 0, \qquad [4.76]$$

$$\begin{aligned}&QV_2(u,x) + a(t)Q_1(x)V_1(u,x) + a(t)Q_{21}(x)V(u)\\ &\qquad + \frac{1}{2}a^2(t)Q_{22}(x)V(u) = \boldsymbol{L}_tV(u).\end{aligned} \qquad [4.77]$$

When the function $V(u)$ depends on the variable x, i.e. it belongs to the space N_Q of zeros for the operator Q, the condition [4.75] holds true.

In accordance with the balance condition and the representation [4.65], we obtain

$$\begin{aligned}&\Pi Q_1(x)V(u) = \Pi q(x)C_0(u,x)V'(u)\\ &= V'(u)\int_X \pi(dx)q(x)C_0(u,x) = qV'(u)\int_X \rho(dx)C_0(u,x) = 0.\end{aligned}$$

The last relation gives an opportunity to present the solution of equation [4.76] in the form

$$V_1(u,x) = R_0\overline{\boldsymbol{C}}_0(x)V(u), \qquad [4.78]$$

where

$$\overline{\boldsymbol{C}}_0(x)V(u) := q(x)C_0(u,x)V'(u). \qquad [4.79]$$

From [4.77], we obtain the representation

$$QV_2(u,x) + \boldsymbol{L}_t(x)V(u) = \boldsymbol{L}_t V(u), \tag{4.80}$$

where

$$\boldsymbol{L}_t(x)V(u) = a(t)Q_1(x)V_1(u,x) + a(t)Q_{21}(x)V(u) + \frac{1}{2}a^2(t)Q_{22}(x)V(u). \tag{4.81}$$

Considering [4.65] and [4.78], we calculate the first summand in [4.81].

$$\begin{aligned} Q_1(x)V_1(u,x) &= C_0(u,x)Q_0 a(t)R_0[\overline{\boldsymbol{C}}_0(x)V(u)]' \\ &= a(t)\overline{\boldsymbol{C}}_0(x)\boldsymbol{P}R_0[\overline{\boldsymbol{C}}_0(x)V(u)]. \end{aligned} \tag{4.82}$$

Here, we use the representation

$$\boldsymbol{P}R_0 = R_0 + g(x)[\Pi - I].$$

Therefore, considering [4.79], for [4.82], we have

$$\boldsymbol{P}R_0\overline{\boldsymbol{C}}_0(x)V(u) = [R_0\overline{\boldsymbol{C}}_0(x) + g(x)\Pi\overline{\boldsymbol{C}}_0(x) - g(x)\overline{\boldsymbol{C}}_0(x)]V(u). \tag{4.83}$$

From the balance condition [4.41], we have

$$\begin{aligned} \Pi\overline{\boldsymbol{C}}_0(x) &= \Pi q(x)\boldsymbol{C}_0(x) \\ &= \int_X \pi(dx)q(x)C_0(u,x) = q\int_X \rho(dx)C_0(u,x) = 0. \end{aligned}$$

By using the last expression and [4.83], from [4.82], we obtain

$$Q_1(x)V_1(u,x) = a(t)[\overline{\boldsymbol{C}}_0(x)R_0\overline{\boldsymbol{C}}_0(x) - \boldsymbol{C}_0(x)\overline{\boldsymbol{C}}_0(x)]V(u). \tag{4.84}$$

For the second summand from [4.81], we use [4.66]:

$$Q_{21}(x)V(u) = C(u,x)q(x)V'(u) = \boldsymbol{C}(x)V(u), \tag{4.85}$$

where

$$\boldsymbol{C}(x)V(u) := q(x)C(u,x)V'(u).$$

For the third summand, we apply [4.68]:

$$Q_{22}(x)V(u) = q(x)C_0^2(u,x)V''(u) = \boldsymbol{C}_0(x)\overline{\boldsymbol{C}}_0(x)V(u). \tag{4.86}$$

In accordance with the representations [4.84] and [4.86], for [4.81], we obtain

$$\boldsymbol{L}_t(x)V(u) = a(t)\boldsymbol{C}(x)V(u) + a^2(t)\boldsymbol{L}_0(x)V(u), \tag{4.87}$$

where

$$\boldsymbol{L}_0(x)V(u) = \overline{\boldsymbol{C}}_0(x)R_0\overline{\boldsymbol{C}}_0(x) - \frac{1}{2}\boldsymbol{C}_0(x)\overline{\boldsymbol{C}}_0(x). \tag{4.88}$$

By averaging the operator $\boldsymbol{L}_t(x)$ at the stationary distribution $\pi(dx)$, we have the representation of the limit operator in the form $\boldsymbol{L}_t = \Pi\boldsymbol{L}_t(x)\Pi$, i.e.

$$\boldsymbol{L}_tV(u) = a(t)\Pi\boldsymbol{C}(x)\Pi V(u) + a^2(t)\Pi\boldsymbol{L}_0(x)\Pi V(u). \tag{4.89}$$

By calculating of right-hand side of [4.89], we obtain for the first summand

$$\Pi\boldsymbol{C}(x)\Pi V(u) = q\int_X \rho(dx)C(u,x)V'(u) = \boldsymbol{C}V(u), \tag{4.90}$$

where

$$\boldsymbol{C}V(u) := C(u)V'(u),$$

and then for the second summand

$$\Pi\boldsymbol{L}_0(x)\Pi V(u) = \Pi\overline{\boldsymbol{C}}_0(x)R_0\overline{\boldsymbol{C}}_0(x)V(u) - \frac{1}{2}\Pi\boldsymbol{C}_0(x)\overline{\boldsymbol{C}}_0(x)V(u). \tag{4.91}$$

For the first summand, from [4.91], we have

$$\begin{aligned}\Pi\overline{\boldsymbol{C}}_0(x)R_0\overline{\boldsymbol{C}}_0(x)V(u) &= \Pi C_0(u,x)q(x)R_0q(x)C_0'(u,x)V'(u)\\ &\quad + \Pi C_0(u,x)q(x)R_0q(x)C_0(u,x)V''(u).\end{aligned}$$

For the second summand [4.91], we obtain

$$\boldsymbol{C}_0(x)\overline{\boldsymbol{C}}_0(x)V(u) = \Pi C_0(u,x)q(x)C_0(u,x)V''(u).$$

Therefore, for [4.91], we have

$$\Pi\boldsymbol{L}_0(x)\Pi V(u) = \Pi C_0(u,x)q(x)R_0q(x)C_0'(u,x)V'(u)$$

$$+\frac{1}{2}[2\Pi C_0(u,x)q(x)R_0q(x)C_0(u,x) - \Pi C_0(u,x)q(x)C_0(u,x)]V''(u).$$

According to [4.90] and the last expression, we obtain for [4.89]

$$\boldsymbol{L}_tV(u) = a(t)\boldsymbol{C}V(u) + a^2(t)\boldsymbol{L}_0V(u), \tag{4.92}$$

where $\boldsymbol{L}_0V(u)$ has the form [4.71].

Furthermore, we use [4.92] for solving equation [4.80] that we can give in the form

$$QV_2(u,x) = [\boldsymbol{L}_t - \boldsymbol{L}_t(x)]V(u).$$

Hence

$$V_2(u,x) = R_0\widetilde{\boldsymbol{L}}_t(x)V(u), \qquad [4.93]$$

where $\widetilde{\boldsymbol{L}}_t(x) := \boldsymbol{L}_t(x) - \boldsymbol{L}_t$.

We highlight the multiplier $a(t)$ in the representation [4.93]. For this purpose, we calculate the centered generator $\widetilde{\boldsymbol{L}}_t(x)$, in accordance with [4.87], [4.88] and [4.92]

$$\begin{aligned}\widetilde{\boldsymbol{L}}_t(x) &= a(t)\boldsymbol{C}(x)V(u) + a^2(t)\boldsymbol{L}_0(x)V(u) - a(t)\boldsymbol{C}V(u) + a^2(t)\boldsymbol{L}_0V(u)\\ &= a(t)[\widetilde{\boldsymbol{C}}(x) + a(t)\widetilde{\boldsymbol{L}}_0(x)]V(u),\end{aligned}$$

where

$$\widetilde{\boldsymbol{C}}(x) = \boldsymbol{C}(u) - C,$$

$$\widetilde{\boldsymbol{L}}_0(x) = \boldsymbol{L}_0(x) - \boldsymbol{L}_0.$$

Consequently, the representation [4.93] has the form

$$V_2(u,x) = R_0[\widetilde{\boldsymbol{C}}(x) + a(t)\widetilde{\boldsymbol{L}}_0(x)]V(u). \qquad [4.94]$$

We use the relations [4.78] and [4.94] to analyze the remainder term [4.74]. We also note that in the relation [4.94], the perturbed Lyapunov function $V(u)$ can be expressed through the first, second and third derivatives that function through the first and second derivatives of the velocities $C(u,x)$ and $C_0(u,x)$.

The same conclusion can be drawn for the representation [4.78].

Using [4.94] and [4.65], for the first summand from [4.74], we have the representation

$$a(t)Q_1(x)V_2(u,x) = a^2(t)\theta_1^{\varepsilon}(x)V(u), \qquad [4.95]$$

where

$$\theta_1^{\varepsilon}(x)V(u) = \overline{\boldsymbol{C}}_0(x)R_0[\widetilde{\boldsymbol{C}}(x) + a(t)\widetilde{\boldsymbol{L}}_0(x)]V(u).$$

For the second summand in [4.74], we use [4.78] and [4.66]

$$a(t)Q_{21}(x)V_1(u,x) = a^2(t)\theta_2^{\varepsilon}(x)V(u), \qquad [4.96]$$

where

$$\theta_2^\varepsilon(x)V(u) = \overline{\boldsymbol{C}}(x)R_0\overline{\boldsymbol{C}}_0(x)V(u).$$

For the third summand, we use [4.68] and [4.78]

$$a^2(t)Q_{22}(x)V_1(u,x) = a^3(t)\theta_3^\varepsilon(x)V(u), \qquad [4.97]$$

where

$$\theta_3^\varepsilon(x)V(u) = q(x)C^2(u,x)R_0[\overline{\boldsymbol{C}}_0(x)V(u)]''.$$

For the fourth summand, we use [4.66] and [4.94]

$$a(t)Q_{21}(x)V_2(u,x) = a^2(t)\theta_4^\varepsilon(x)V(u), \qquad [4.98]$$

where

$$\theta_4^\varepsilon(x)V(u) = \overline{\boldsymbol{C}}(x)R_0[\widetilde{\boldsymbol{C}}(x) + a(t)\widetilde{\boldsymbol{L}}_0(x)]V(u).$$

For the fifth summand, we use [4.68] and [4.94]

$$a^2(t)Q_{22}(x)V_2(u,x) = a^3(t)\theta_5^\varepsilon(x)V(u), \qquad [4.99]$$

where

$$\theta_5^\varepsilon(x)V(u) = q(x)C^2(u,x)R_0[\widetilde{\boldsymbol{C}}(x)V(u) + a(t)\widetilde{\boldsymbol{L}}_0(x)V(u)]''.$$

Using [4.95]–[4.99], the remainder term [4.74] can be represented in the form

$$\theta_{t0}^\varepsilon(x)V(u) = a^2(t)\widehat{\theta}_{t0}^\varepsilon(x)V(u), \qquad [4.100]$$

where

$$\begin{aligned}\widehat{\theta}_{t0}^\varepsilon(x)V(u) = \theta_1^\varepsilon(x)V(u) &+ \theta_2^\varepsilon(x)V(u)\\ &+ \frac{1}{2}a(t)\theta_3^\varepsilon(x)V(u) + \varepsilon\theta_4^\varepsilon(x)V(u)\\ &+ \varepsilon\frac{1}{2}a(t)\theta_5^\varepsilon(x)V(u).\end{aligned}$$

Therefore, in accordance with [4.73], [4.77], [4.92] and [4.98], we obtain the asymptotic representation

$$\boldsymbol{L}_{t0}V^\varepsilon(u,x) = a(t)\boldsymbol{C}V(u) + a^2(t)\boldsymbol{L}_0V(u) + \varepsilon a^2(t)\widehat{\theta}_{t0}^\varepsilon(x)V(u).$$

Step 2. Considering the form of the truncated operator $\boldsymbol{L}_{t0}$ in [4.72], for the whole generator $\boldsymbol{L}_t^\varepsilon$, we have the representation

$$\boldsymbol{L}_t^\varepsilon = \boldsymbol{L}_{t0} + \varepsilon a^2(t)\boldsymbol{G}_t^\varepsilon(x).$$

Thus, the singular perturbation problem for the generator $\boldsymbol{L}_t^\varepsilon$ on the functions $V^\varepsilon(u,x)$ has the form

$$\boldsymbol{L}_t^\varepsilon V^\varepsilon(u,x) = \boldsymbol{L}_{t0} V^\varepsilon(u,x) + \varepsilon a^2(t)\boldsymbol{G}_t^\varepsilon(x)[V(u) + \varepsilon V_1(u,x) + \varepsilon^2 V_2(u,x)].$$

Using the representation [4.98], from [4.100], we obtain

$$\boldsymbol{L}_t^\varepsilon V^\varepsilon(u,x) = a(t)\boldsymbol{C}V(u) + a^2(t)\boldsymbol{L}_0 V(u) + \varepsilon a^2(t)\boldsymbol{H}_t^\varepsilon(x)V(u), \qquad [4.101]$$

where

$$\boldsymbol{H}_t^\varepsilon(x)V(u) = \widehat{\theta}_{t0}^\varepsilon(x)V(u) + \boldsymbol{G}_t^\varepsilon(x)V(u) + \varepsilon a(t)\boldsymbol{G}_t^\varepsilon(x)R_0\overline{\boldsymbol{C}}_0(x)V(u)$$
$$+ \varepsilon^2 a(t)\boldsymbol{G}_t^\varepsilon(x)R_0[\widetilde{\boldsymbol{C}}(x) + a(t)\widetilde{\boldsymbol{L}}_0(x)]V(u). \qquad [4.102]$$

In conclusion, we note that from the representations [4.98] and [4.68], we obtain the boundedness right-hand side of [4.102], i.e. the relation

$$\| \boldsymbol{H}_t^\varepsilon(x)V(u) \| \leq M, M > 0$$

holds true.

LEMMA 4.8.– *For the operators $\boldsymbol{L}_0$ and $\boldsymbol{H}_t^\varepsilon$ from the representation [4.70], we have the estimates*

$$|\boldsymbol{L}_0 V(u)| \leq c(1 + V(u)), \qquad [4.103a]$$

$$|\boldsymbol{H}_t^\varepsilon V(u)| \leq c(1 + V(u)). \qquad [4.103b]$$

PROOF.– We use the boundedness of the potential R_0 and the boundedness of the intensity $q(x)$, and then we estimate the summand in [4.74] by the functions $w_k(u)$, which are determined by recursion equations [4.50] and [4.51]. The first and second summands in [4.74] are the major functions $w_1(u)$ and $w_2(u)$, respectively. Thus, we have the estimate [4.103a] according to condition 2) of the theorem at $k = 1, 2$.

We consider the representation [4.102] for the justification of the estimate [4.103b]. The first summand is majorated by the functions $w_1(u)$, $w_2(u)$ and $w_3(u)$. Then, under condition C2 of the theorem, the estimate

$$|\widehat{\theta}_{t0}^\varepsilon(x)V(u)| \leq c(1 + V(u))$$

holds true.

According to [4.68], the second summand is majorated by the functions $w_1(u)$, $w_2(u)$ and $w_3(u)$, then we have an estimate

$$|\boldsymbol{G}_t^\varepsilon(x)V(u)| \leq c(1 + V(u)).$$

For the third summand, we additionally use the boundedness of the function $a(t)$, and the function $w_4(u)$, obtaining

$$\varepsilon|a(t)\boldsymbol{G}_t^\varepsilon(x)R_0\overline{\boldsymbol{C}}_0(x)V(u)| \leq c(1+V(u)).$$

For the fourth summand, using additionally the function $w_4(u)$, we have

$$\varepsilon^2|a(t)\boldsymbol{G}_t^\varepsilon(x)R_0[\widetilde{\boldsymbol{C}}(x)+a(t)\widetilde{\boldsymbol{L}}_0(x)]V(u)| \leq c(1+V(u)).$$

As a result, we obtain [4.103b].

We note that under the conditions of the theorem, the estimate

$$\mathbf{L}_t^\varepsilon V(u) \leq -\delta a(t)V(u), \delta > 0$$

holds true.

PROOF OF THEOREM.– Since for the generator $\boldsymbol{L}_t^\varepsilon$ estimate [4.102] holds true, it is enough to use the lemma by Nevelson and Khasminsky (1972) to complete the proof of theorem.

Conclusion. During the investigation of the SAP in a random Markov environment, it is first necessary to provide a statistical analysis of the Markov environment, i.e. to find a stationary distribution of the perturbed Markov process that defines the random environment and then to find the averaging regression function $C(u)$ that determines the equilibrium point of the SAP.

4.2. Semi-Markov environment

4.2.1. *SAP under the averaging scheme*

Jumping SAP can be given by the relation (we assume that $\sum\limits_{k=0}^{-1} a_k^\varepsilon C(u_k^\varepsilon, x_k^\varepsilon) = 0$)

$$u^\varepsilon(t) = u + \varepsilon \sum_{k=0}^{\nu(t/\varepsilon)-1} a_k^\varepsilon C(u_k^\varepsilon, x_k^\varepsilon), u^\varepsilon(0) = u, \qquad [4.104]$$

where the regression function $C(u,x) = (C_k(u,x), k = \overline{1,d})$, $u \in R^d, x \in X$, satisfies the conditions of the existing global solution for accompanying systems

$$du_x(t)/dt = C(u_x(t), x), x \in X.$$

The semi-Markov process $x(t), t \geq 0$, on a standard phase space $(X, \boldsymbol{X})$ (Korolyuk and Limnios 2005) is given by semi-Markov kernel

$$Q(x,B,t) = P(x,B)G_x(t), x \in X, B \in \boldsymbol{X}, t \geq 0.$$

Stochastic kernel $P(x, B)$ is defined by transition probabilities of embedded Markov chain $x_n, n \geq 0$,

$$P(x, B) := P\{x_{n+1} \in B \,|x_n = x\},$$

and $G_x(t)$, $x \in X, t \geq 0$, is the distributing staying time function θ_x at state $x \in X$:

$$G_x(t) := P\{\theta_{n+1} \leq t|x_n = x\}.$$

In [4.104], the normalizing sequence a_n^ε, $n \geq 0$, is defined by the function value $a(t), t \geq 0$ through the relation:

$$a_n^\varepsilon := a(\tau_n^\varepsilon), \tau_n^\varepsilon := \varepsilon\tau_n, n \geq 0, \qquad [4.105]$$

where $\tau_n, n \geq 0$, are moments of Markov renewals of the semi-Markov process $x(t), t \geq 0$, with the counting process $\nu(t) := \max\{n : \tau_n \leq t\}, t \geq 0$.

From the semi-Markov process $x(t), t \geq 0$, we consider the accompanying Markov process $x_0(t), t \geq 0$, which is defined by the generator

$$Q\varphi(x) = q(x) \int_X P(x, dy) \left[\varphi(y) - \varphi(x)\right]. \qquad [4.106]$$

From [4.105], we have embeddings of the discrete SAP $u_n^\varepsilon, n \geq 0$, into jumping SAP [4.104]:

$$u_n^\varepsilon = u(\tau_n^\varepsilon), x_n^\varepsilon = x(\tau_n^\varepsilon), n \geq 0. \qquad [4.107]$$

We investigate the convergence of SAP [4.104] under the conditions of exponential stability of the averaging system

$$\frac{du(t)}{dt} = C(u(t)), C(u) = q \int_X \rho(dx)C(u, x). \qquad [4.108]$$

Henceforth, without limitation on the generality, we assume $u_0 = 0$, i.e.

$$C(0) = 0.$$

THEOREM 4.3.– *We suppose that the Lyapunov function $V(u) \in C^3(R^d), u \in R^d$, for the averaging system [4.108] is satisfied under the following conditions:*

$C1$: *the exponential stability of the averaged system [4.108] is fulfilled*

$$C(u)V'(u) \leq -c_0 V(u), c_0 > 0;$$

C2 : for $\overline{C}(u) := \max_{x\in X} C(u,x)$*, the following estimates hold true:*

$$\left|\overline{C}(u)V'(u)\right| \le c_1(1+V(u)), c_1 > 0,$$

$$\left|\overline{C}(u)[\overline{C}(u)V'(u)]'\right| \le c_2(1+V(u)), c_2 > 0,$$

$$\left|\overline{C}(u)\overline{C}(u)[\overline{C}(u)V'(u)]''\right| \le c_3(1+V(u)), c_3 > 0.$$

Finally, the normalizing function $a(t)$ *is monotonically decreasing, limited and satisfies the following conditions:*

$$C3 : \int_0^\infty a(t)dt = \infty, \int_0^\infty a^2(t)dt < \infty.$$

Then, at each positive $\varepsilon \le \varepsilon_0$*, where* ε_0 *is small enough, SAP [4.104] converges with probability 1 to a unique equilibrium point of the averaged evolutionary system [4.108]:*

$$P\{\lim_{t\to\infty} u^\varepsilon(t) = 0\} = 1.$$

For jumping SAP [4.104], we consider the extended renewal Markov process

$$u_n^\varepsilon = u^\varepsilon(\tau_n^\varepsilon), x_n^\varepsilon = x(\tau_n^\varepsilon), \tau_n^\varepsilon = \varepsilon\tau_n, n \ge 0. \qquad [4.109]$$

The compensating operator for the process [4.109] is defined by the conditional expectation

$$\begin{aligned}\boldsymbol{L}_t^\varepsilon\varphi(u,x,t) := \varepsilon^{-1}q(x)E[\varphi(u_{n+1}^\varepsilon, x_{n+1}^\varepsilon, \tau_{n+1}^\varepsilon)\\ - \varphi(u,x,t)\,|u_n^\varepsilon = u,\; x_n^\varepsilon = x, \tau_n^\varepsilon = t].\end{aligned} \qquad [4.110]$$

LEMMA 4.9.– *The compensating operator [4.110] on the test functions* $\varphi(u,x)$*,* $u \in R^d$*,* $x \in X$*, has the representation*

$$\boldsymbol{L}_t^\varepsilon\varphi(u,x) := \varepsilon^{-1}Q\varphi(u,x) + \varepsilon^{-1}q(x)\boldsymbol{P}\boldsymbol{C}_t^\varepsilon(x)\varphi(u,x), \qquad [4.111]$$

where

$$\boldsymbol{C}_t^\varepsilon(x)\varphi(u,x) = \varphi(u+\varepsilon a(t)C(u,x),x) - \varphi(u,x), \qquad [4.112]$$

and the operator ***P*** *is given by the relation*

$$\boldsymbol{P}\varphi(\cdot,x) := \int_X P(x,dy)\varphi(\cdot,y). \qquad [4.113]$$

PROOF.– We calculate the conditional expectation

$$\begin{aligned} E[\varphi(u^\varepsilon_{n+1}, x^\varepsilon_{n+1})|u^\varepsilon_n = u, x^\varepsilon_n = x, \tau^\varepsilon_n = t] &= E_{u,x,t}\varphi(u^\varepsilon_{n+1}, x^\varepsilon_{n+1}) \\ &= E_{u,x,t}\varphi(u^\varepsilon_n + \Delta u^\varepsilon_{n+1}, x^\varepsilon_{n+1}), \end{aligned} \quad [4.114]$$

where $\Delta u^\varepsilon_{n+1} := u^\varepsilon_{n+1} - u^\varepsilon_n$. To calculate the increment $\Delta u^\varepsilon_{n+1}$, we use the following representations from [4.104] and [4.109], i.e.

$$u^\varepsilon_{n+1} := u^\varepsilon(\tau^\varepsilon_{n+1}) = u + \varepsilon \sum_{k=0}^{n} a^\varepsilon_k C(u^\varepsilon_k, x^\varepsilon_k),$$

$$u^\varepsilon_n := u^\varepsilon(\tau^\varepsilon_n) = u + \varepsilon \sum_{k=0}^{n-1} a^\varepsilon_k C(u^\varepsilon_k, x^\varepsilon_k).$$

and therefore we have

$$\Delta u^\varepsilon_{n+1} = \varepsilon a^\varepsilon_n C(u^\varepsilon_n, x^\varepsilon_n). \quad [4.115]$$

Substituting [4.115] into [4.110], we obtain

$$E_{u,x,t}\varphi(u^\varepsilon_n + \Delta u^\varepsilon_{n+1}, x^\varepsilon_{n+1}) = E_{u,x,t}\varphi(u + \varepsilon a(t)C(u, x), y),$$

where $y = x^\varepsilon_{n+1}$ is the value of the process $x(t), t \geq 0$, at the next jumping moment (see [4.107]).

Thus, the compensating operator [4.110] has the form

$$\begin{aligned} \boldsymbol{L}^\varepsilon_t \varphi(u, x) :&= \varepsilon^{-1} q(x) E_{u,x,t}[\varphi(u + \varepsilon a(t)C(u, x), y) - \varphi(u, x)] \\ &= \varepsilon^{-1} q(x) \int_X P(x, dy)[\varphi(u + \varepsilon a(t)C(u, x), y) - \varphi(u, x)]. \end{aligned}$$

Using summand $\pm\varphi(u, y)$, considering the last equality, we have

$$\begin{aligned} \boldsymbol{L}^\varepsilon_t \varphi(u, x) &= \varepsilon^{-1} q(x) \int_X P(x, dy)[\varphi(u, y) - \varphi(u, x)] \\ &+ \varepsilon^{-1} q(x) \int_X P(x, dy)[\varphi(u + \varepsilon a(t)C(u, x), y) - \varphi(u, y)]. \end{aligned} \quad [4.116]$$

According to the definition of the generator Q, from [4.116], we obtain [4.111] and the lemma is proved.

CONSEQUENCE 4.1.– The compensating operator [4.111] on the test functions $\varphi(u,\cdot) \in C^3(R^d)$ under conditions C2 of the theorem has the asymptotic representation

$$\mathbf{L}_t^\varepsilon \varphi(u,x) = \varepsilon^{-1} Q\varphi(u,x) + a(t)Q_1(x)\varphi(u,x) + \varepsilon a^2(t)\theta_1^\varepsilon(x)\varphi(u,x), \quad [4.117]$$

where

$$Q_1(x)\varphi(u,x) = C(u,x)Q_0(x)\varphi'_u(u,x), \quad [4.118]$$

and

$$Q_0\varphi(u,x) = q(x)\mathbf{P}\varphi(u,x),$$

and the remainder term $\theta_1^\varepsilon(x)\varphi(u,x) = \frac{1}{2}C^2(u,x)Q_0\varphi''_u(\theta u,x), 0 \le \theta \le 1$ is that

$$\varepsilon a^2(t)\left\|\theta_1^\varepsilon(x)\varphi(u,x,t)\right\| \to 0, \varepsilon \to 0.$$

PROOF.– We have the decomposition

$$\boldsymbol{C}_t^\varepsilon(x)\varphi(u,x) = \varepsilon a(t)C(u,x)\varphi'_u(u,x) + \theta_C^\varepsilon(x)\varphi(u,x), \quad [4.119]$$

for the operator $\boldsymbol{C}_t^\varepsilon(x)$ from [4.113] in accordance with the smoothness of the function $\varphi(u,x)$, where

$$\theta_C^\varepsilon(x)\varphi(u,x) = \varepsilon^2 a^2(t)\frac{1}{2}C^2(u,x)\varphi''_u(\theta u,x), 0 \le \theta \le 1.$$

We substitute decomposition [4.119] into [4.111] and then make a series of transformations

$$\boldsymbol{L}_t^\varepsilon \varphi(u,x) = \varepsilon^{-1}Q\varphi(u,x) + \varepsilon^{-1}q(x)\varepsilon a(t)C(u,x)\boldsymbol{P}\varphi'_u(u,x)$$
$$+ \varepsilon^{-1}q(x)\varepsilon^2 a^2(t)\frac{1}{2}C^2(u,x)\boldsymbol{P}\varphi''_u(\theta u,x).$$

Therefore, we have the representation [4.117] with the remainder term $\theta_1^\varepsilon(x)$ $\varphi(u,x)$.

In accordance with the properties of the functions $\varphi(u,x)$, $a(t)$ and $C(u,x)$, we also obtain

$$\varepsilon a^2(t)\left\|\theta_1^\varepsilon(x)\varphi(u,x)\right\| \to 0, \varepsilon \to 0$$

and the statement of consequence is proved.

The first step of proving the theorem is solving the singular perturbation problem for the operator [4.117] on the perturbed Lyapunov function

$$V^\varepsilon(u,x) = V(u) + \varepsilon a(t)V_1(u,x), \quad [4.120]$$

where $V(u)$ is the Lyapunov function for the system [4.108].

LEMMA 4.10.– *The compensating operator* $\boldsymbol{L}_t^\varepsilon$*, under condition C2 of the theorem on the functions* $V^\varepsilon(u,x)$*, where* $V(u) \in C^3(R^d)$*, has the asymptotic representation*

$$\boldsymbol{L}_t^\varepsilon V^\varepsilon(u,x,t) = a(t)C(u)V'(u) + \varepsilon a^2(t)\theta_L^\varepsilon(x)V(u), \qquad [4.121]$$

where

$$\theta_L^\varepsilon(x)V(u) = \theta_0^\varepsilon(x)V(u) + \theta_1^\varepsilon(x)V(u),$$

$$\begin{aligned}\theta_0^\varepsilon(x)V(u) &= C(u,x)Q_0R_0\widetilde{C}'_u(u,x)V'(u)\\ &\quad + C(u,x)[q(x)R_0 - I]\widetilde{C}(u,x)V''(u),\end{aligned}$$

$$\theta_1^\varepsilon(x)V(u) = \frac{1}{2}C^2(u,x)q(x)V''(\theta u), 0 \le \theta \le 1,$$

$$\widetilde{C}(u,x) = q(x)C(u,x) - C(u).$$

PROOF.– First, we consider the solution of the singular perturbation problem for the truncated operator for [4.117], i.e.

$$\boldsymbol{L}_{t0}^\varepsilon = \varepsilon^{-1}Q + a(t)Q_1(x),$$

The decomposition of the operator $\boldsymbol{L}_{t0}^\varepsilon$ on the function $V^\varepsilon(u,x)$ has the form

$$\begin{aligned}&\boldsymbol{L}_{t0}^\varepsilon V^\varepsilon(u,x) = [\varepsilon^{-1}Q + a(t)Q_1(x)][V(u) + \varepsilon a(t)V_1(u,x)]\\ &= \varepsilon^{-1}QV(u) + a(t)[QV_1(u,x) + Q_1(x)V(u)] + \varepsilon a^2(t)Q_1(x)V_1(u,x). \qquad [4.122]\end{aligned}$$

From the solvability condition for decomposition [4.122], we have

$$QV_1(u,x) + Q_1(x)V(u) = \widehat{Q}_1V(u) \qquad [4.123]$$

where

$$\widehat{Q}_1\Pi = \Pi Q_1(x)\Pi. \qquad [4.124]$$

We calculate the right-hand side of [4.124]. From the representation [4.118] and the action of the projector Π, we have

$$\begin{aligned}\Pi Q_1(x)\Pi &= \Pi C(u,x)Q_0(x)\Pi = \Pi C(u,x)q(x)\boldsymbol{P}\Pi\\ &= \Pi q(x)C(u,x)\Pi = \int_X \pi(dx)q(x)C(u,x)\\ &= q\int_X \rho(dx)C(u,x) = C(u).\end{aligned}$$

Consequently, considering the derivative calculation in [4.118], we obtain for the right-hand side of [4.123]

$$\widehat{Q}_1 V(u) = C(u)V'(u). \qquad [4.125]$$

Therefore, from [4.123], we have

$$QV_1(u,x) = a(t)(\widehat{Q}_1 - Q_1(x))V(u).$$

According to potential property R_0 from the last equality, we have solution

$$V_1(u,x) = R_0\tilde{Q}_1(x)V(u), \qquad [4.126]$$

where

$$\tilde{Q}_1(x) = Q_1(x) - \widehat{Q}_1. \qquad [4.127]$$

Then, we calculate the right-hand side of [4.127] considering [4.118] and [4.125],

$$Q_1(x) - \widehat{Q}_1 = [C(u,x)q(x)\boldsymbol{P} - C(u)] = \widetilde{C}(u,x).$$

Thus, for the perturbation $V_1(u,x)$ of the Lyapunov function $V(u)$ in [4.126], we obtain

$$V_1(u,x) = R_0\widetilde{\mathbf{C}}(x)V(u), \qquad [4.128]$$

where

$$\widetilde{\mathbf{C}}(x)V(u) = \tilde{C}(u,x)V'(u).$$

We use [4.128] to calculate the last summand in [4.122]. For this summand, we consider multiplier $\theta_0^\varepsilon(x)$, i.e.

$$\theta_0^\varepsilon(x)V(u) := Q_1(x)V_1(u,x).$$

According to [4.118] and [4.128], we calculate $\theta_0^\varepsilon(x)$

$$\begin{aligned}\theta_0^\varepsilon(x)V(u) &= C(u,x)Q_0[R_0\widetilde{\mathbf{C}}(x)V(u)]' \\ &= C(u,x)Q_0R_0[\widetilde{C}(u,x)V'(u)]' \\ &= q(x)\boldsymbol{P}R_0[\widetilde{C}'_u(u,x)V'(u) + \widetilde{C}(u,x)V''(u)]. \qquad [4.129]\end{aligned}$$

From equality $q(x)\boldsymbol{P}R_0 = q(x)R_0 + \Pi - I$, we have the decomposition [4.129]

$$\begin{aligned}\theta_0^\varepsilon(x)V(u) &= q(x)\boldsymbol{P}R_0[\widetilde{C}'_u(u,x)V'(u) + \widetilde{C}(u,x)V''(u)] \\ &= [q(x)R_0 + \Pi - I][\widetilde{C}'_u(u,x)V'(u) + \widetilde{C}(u,x)V''(u)] \\ &= q(x)\boldsymbol{P}R_0\widetilde{C}'_u(u,x)V'(u) + q(x)R_0\widetilde{C}(u,x)V''(u) \\ &\quad + \Pi\widetilde{C}(u,x)V''(u) - \widetilde{C}(u,x)V''(u) \\ &= q(x)\boldsymbol{P}R_0\widetilde{C}'_u(u,x)V'(u) + [q(x)R_0 - I]\widetilde{C}(u,x)V''(u),\end{aligned}$$

since

$$\begin{aligned}\Pi\widetilde{C}(u,x) &= \Pi[q(x)C(u,x) - C(u)] \\ &= \int_X \pi(dx)q(x)C(u,x) - C(u)\int_X \pi(dx)\boldsymbol{1}(X) \\ &= q\int_X \rho(dx)C(u,x) - C(u) = C(u) - C(u) = 0.\end{aligned}$$

Therefore, we have the representation for the remainder term $\theta_0^\varepsilon(x)V(u)$.

Finally, the right-hand side of [4.122] gives the representation

$$\boldsymbol{L}_{t0}^\varepsilon V^\varepsilon(u,x) = a(t)C(u)V'(u) + \varepsilon a^2(t)\theta_0^\varepsilon(x)V(u). \qquad [4.130]$$

Returning to the full view of the compensating operator [4.117], we represent it in the form

$$\boldsymbol{L}_t^\varepsilon = \boldsymbol{L}_{t0}^\varepsilon + \varepsilon a^2(t)\theta_1^\varepsilon(x).$$

Consequently, we obtain decomposition on functions [4.120]:

$$\boldsymbol{L}_t^\varepsilon V^\varepsilon(u,x) = [\,\boldsymbol{L}_{t0}^\varepsilon + \varepsilon a^2(t)\theta_1^\varepsilon(x)][V(u) + \varepsilon a(t)V_1(u,x)], \qquad [4.131]$$

where $V_1(u,x)$ is calculated by formula [4.128].

Considering [4.130], we transform the right-hand side of [4.131]:

$$\begin{aligned}\boldsymbol{L}_t^\varepsilon V^\varepsilon(u,x) &= \boldsymbol{L}_{t0}^\varepsilon V^\varepsilon(u,x) + \varepsilon a^2(t)\theta_1^\varepsilon(x)[V(u) + \varepsilon a(t)V_1(u,x)] \\ &= a(t)C(u)V'(u) + \varepsilon a^2(t)\theta_0^\varepsilon(x)V(u) \\ &\quad + \varepsilon a^2(t)\theta_1^\varepsilon(x)[V(u) + \varepsilon a(t)R_0\widetilde{C}(u,x)V'(u)]. \qquad [4.132]\end{aligned}$$

Furthermore, we calculate the last summand from [4.132]:

$$\begin{aligned}\varepsilon a^2(t)\theta_1^\varepsilon(x)[V(u) + \varepsilon a(t)R_0\widetilde{C}(u,x)V'(u)] &= \varepsilon a^2(t)\theta_1^\varepsilon(x)V(u) \\ &\quad + \varepsilon^2 a^3(t)\theta_1^\varepsilon(x)R_0[\widetilde{C}(u,x)V'(u)].\end{aligned}$$

Calculating the right-hand side at the obtained decomposition, we have

$$\theta_1^\varepsilon(x)V(u) = \frac{1}{2}C^2(u,x)q(x)V''(\theta u), 0 \le \theta \le 1,$$

$$\theta_1^\varepsilon(x)R_0[\widetilde{C}(u,x)V'(u)] = \frac{1}{2}C^2(u,x)Q_0R_0[\widetilde{C}(u,x)V'(u)]_u''|_{u=\theta u}.$$

Under condition C2 of the theorem and the boundedness of the operators Q_0 and R_0, we obtain

$$||\theta_1^\varepsilon(x)R_0[\widetilde{C}(u,x)V'(u)]|| \leq M, M > 0, x \in X.$$

Therefore, in accordance with the second order of smallness of the variable ε, we can ignore the second term in decomposition of the remainder term.

As a result, from [4.130], we obtain [4.131] and the lemma is proved.

PROOF OF THE THEOREM.– We make the proof in several steps.

First of all, we establish the key inequality.

LEMMA 4.11.– *The compensating operator $\boldsymbol{L}_t^\varepsilon$ on the perturbed Lyapunov function [4.130] under the conditions of the theorem has the estimate*

$$\boldsymbol{L}_t^\varepsilon V^\varepsilon(u,x) \leq -\delta a(t)V(u), \delta > 0. \quad [4.133]$$

PROOF.– Using estimates $C2$ for the remainder term $\theta_L^\varepsilon(x)V(u)$, condition $C1$ of exponential stability for the system [4.108], as well as monotonicity and boundedness of the normalizing function $a(t)$, we have

$$\begin{aligned}\boldsymbol{L}_t^\varepsilon V^\varepsilon(u,x) &= a(t)C(u)V'(u) + \varepsilon a^2(t)\theta_L^\varepsilon(x)V(u) \\ &\leq -a(t)(\delta^* - \varepsilon c)V(u) \leq -a(t)\delta V(u), \delta > 0. \end{aligned} \quad [4.134]$$

The last inequality holds true at each $\varepsilon \leq \varepsilon_0$, where ε_0 satisfies $\varepsilon_0 < \delta^*/c$.

Second, a clear form for perturbation [4.128] and condition $C2$ of the theorem gives an estimate of the perturbed Lyapunov function

$$0 < (1 - \varepsilon a(t)c)V(u) \leq V^\varepsilon(u,x) \leq (1 + \varepsilon a(t)c)V(u).$$

Third, we prove that the process

$$V_n^\varepsilon := V^\varepsilon(u_n^\varepsilon, x_n^\varepsilon), n \geq 0, \quad [4.135]$$

is non-negative supermartingale. We note that the extended Markov renewal process [4.107] generates on the test functions $\varphi(u,x)$, $u \in R^d$, $x \in X$, martingale

$$\mu_{n+1} = \varphi(u_{n+1}^\varepsilon, x_{n+1}^\varepsilon) - \varphi(u_0, x_0) - \varepsilon \sum_{k=0}^{n} \theta_{k+1} \boldsymbol{L}_{\tau_k^\varepsilon}^\varepsilon \varphi(u_k^\varepsilon, x_k^\varepsilon), \quad [4.136]$$

regarding the flow of σ-algebras $F_n^\varepsilon = \sigma\{u_k^\varepsilon, x_k^\varepsilon, \tau_k^\varepsilon, 0 \leq k \leq n\}, n \geq 0$. Indeed, at notation $\varphi_n^\varepsilon := \varphi(u_n^\varepsilon, x_n^\varepsilon)$, we have

$$E[\mu_{n+1} - \mu_n \,|F_n^\varepsilon] = E_{F_n^\varepsilon}[\varphi_{n+1}^\varepsilon - \varphi_n^\varepsilon] - E_{F_n^\varepsilon}[\varepsilon\theta_{n+1}\boldsymbol{L}_{\tau_n^\varepsilon}^\varepsilon \varphi_n^\varepsilon]. \quad [4.137]$$

Considering the definition of the compensation operator [4.110], we calculate the second summand in [4.137]:

$$E_{F_n^\varepsilon}[\varepsilon\theta_{n+1}\, \boldsymbol{L}^\varepsilon_{\tau_n^\varepsilon}\varphi_n^\varepsilon] = \varepsilon g(x_n^\varepsilon)\boldsymbol{L}^\varepsilon_{\tau_n^\varepsilon}\varphi_n^\varepsilon = \varepsilon g(x_n^\varepsilon)\varepsilon^{-1}q(x_n^\varepsilon)E_{F_n^\varepsilon}[\varphi_{n+1}^\varepsilon - \varphi_n^\varepsilon]$$
$$= E_{F_n^\varepsilon}[\varphi_{n+1}^\varepsilon - \varphi_n^\varepsilon].$$

Therefore, from [4.137], we obtain the martingale property for [4.136]:

$$E[\mu_{n+1} - \mu_n \,|F_n^\varepsilon] = 0, n \geq 0. \qquad [4.138]$$

We use [4.138] to prove the next lemma.

LEMMA 4.12.– *The process [4.135] is non-negative supermartingale.*

PROOF.– We consider in [4.136] the perturbed Lyapunov function $V^\varepsilon(u,x)$ instead of an arbitrary function and then we use the martingale characterization of the process [4.136], i.e.

$$E[V_{n+1}^\varepsilon \,|F_n^\varepsilon] = V_0^\varepsilon + \varepsilon E[\sum_{k=0}^{n} \theta_{k+1}\, \boldsymbol{L}^\varepsilon_{\tau_k^\varepsilon} V_k^\varepsilon] + \mu_n^\varepsilon, \qquad [4.139]$$

since from [4.138], we have $E[\mu_{n+1}^\varepsilon \,|F_n^\varepsilon] = \mu_n^\varepsilon$. We substitute into [4.139] instead of μ_n^ε its value from [4.136]:

$$E[V_{n+1}^\varepsilon \,|F_n^\varepsilon] = V_0^\varepsilon + \varepsilon g(x_n^\varepsilon)\boldsymbol{L}^\varepsilon_{\tau_n^\varepsilon}V_n^\varepsilon + \varepsilon[\sum_{k=0}^{n-1} \theta_{k+1}\boldsymbol{L}^\varepsilon_{\tau_k^\varepsilon}V_k^\varepsilon]$$
$$+ V_n^\varepsilon - V_0^\varepsilon - \varepsilon[\sum_{k=0}^{n-1} \theta_{k+1}\boldsymbol{L}^\varepsilon_{\tau_k^\varepsilon}V_k^\varepsilon].$$

From here, we have

$$E[V_{n+1}^\varepsilon \,|F_n^\varepsilon] = V_n^\varepsilon + \varepsilon g(x_n^\varepsilon)\boldsymbol{L}^\varepsilon_{\tau_n^\varepsilon}V_n^\varepsilon,$$

and considering key inequality [4.133], we obtain $E[V_{n+1}^\varepsilon \,|F_n^\varepsilon] \geq V_n^\varepsilon, n \geq 0$ and the lemma is proved.

Using estimate [4.123] and Lemma 4.12, the proof of the theorem is completed by the scheme of the proof of Theorem 4.1.

4.2.2. *Jumping SAP in the diffusion approximation scheme*

In this section, we investigate the convergence of jumping SAP in a semi-Markov environment using the asymptotic properties of the compensating operator for the extended Markov renewal process and solving its singular perturbation problem (Korolyuk and Limnios 2005).

The jumping stochastic approximation procedure under the diffusion approximation scheme can be given in the form (we assume $\sum_{k=0}^{-1} a_k^\varepsilon C^\varepsilon(u_k^\varepsilon, x_k^\varepsilon) := 0$):

$$u^\varepsilon(t) = u + \varepsilon^2 \sum_{k=0}^{\nu(t/\varepsilon^2)-1} a_k^\varepsilon C^\varepsilon(u_k^\varepsilon, x_k^\varepsilon), u^\varepsilon(0) = u, t > 0. \qquad [4.140]$$

Sequence $a_n^\varepsilon, n \geq 0$, is defined by the value of the function $a(t)$, $t > 0$, using the relation:

$$a_n^\varepsilon = a(\tau_n^\varepsilon), \tau_n^\varepsilon = \varepsilon^2 \tau_n, n \geq 0, \qquad [4.141]$$

where $\tau_n, n \geq 0$, is the Markov renewal moments for a uniformly ergodic semi-Markov process (SMP) $x(t), t \geq 0$, on a standard phase space $(X, \boldsymbol{X})$ with the counting process

$$\nu(t) := \max\{n : \tau_n \leq t\}, t \geq 0.$$

In SAP [4.140], together with [4.141], we have the relation

$$u_n^\varepsilon = u^\varepsilon(\tau_n^\varepsilon), x_n^\varepsilon = x(\tau_n^\varepsilon), \tau_n^\varepsilon := \varepsilon^2 \tau_n, n \geq 0. \qquad [4.142]$$

Function $C^\varepsilon(u, x)$ in SAP [4.140] is that

$$C^\varepsilon(u, x) := C(u, x) + \varepsilon^{-1} C_0(u, x), x \in X, \qquad [4.143]$$

and it satisfies the condition of the existing global solution for accompanying systems

$$du_x(t)/dt = C^\varepsilon(u_x(t), x), x \in X.$$

For the perturbation $C_0(u, x)$ of the regression function, we assume that the balance condition is provided

BC1: $\tilde{\Pi} C_0(u, x) := \int_X \rho(dx) C_0(u, x) = 0,$

We investigate the convergence of jumping SAP [4.143] under the conditions of exponential stability for the averaging system

$$du(t)/dt = C(u(t)), C(u) := q \int_X \rho(dx) C(u, x). \qquad [4.144]$$

We assume without limitation the generality $u_0 = 0$, i.e. the equality

$$C(0) = 0 \qquad [4.145]$$

holds true.

Hereafter, the operator $\boldsymbol{P}$ is defined by the kernel $P(x, B), B \in X$,

$$\boldsymbol{P}\varphi(x) := \int_X P(x, dy)\varphi(y), \qquad [4.146]$$

and R_0 is the potential to the operator Q, where $QR_0 = RQ_0 = \Pi - I$.

Here, we use the conditions of convergence of the solution of the evolutionary equation

$$du_0^\varepsilon(t) = [C(u_0^\varepsilon(t), x(t/\varepsilon^2)) + \varepsilon^{-1}C_0(u_0^\varepsilon(t), x(t/\varepsilon^2))]dt, \qquad [4.147]$$

to the solution of the stochastic differential equation

$$du_0(t) = C(u_0(t))dt + d\zeta(t),$$

$$d\zeta(t) = b(u_0(t))dt + \frac{1}{2}SpB(u_0(t))dw_t.$$

The diffusion process $\zeta(t), t \geq 0$, is defined by the shift function

$$b(u) = b_1(u) - b_2(u), \qquad [4.148]$$

where

$$b_1(u) := \int_X \pi(dx)\overline{C}_0(u, x)R_0\overline{C}_0'(u, x),$$

$$b_2(u) := \frac{1}{2}q\int_X \rho(dx)C_0(u, x)C_0'(u, x),$$

$$\overline{C}_0(u, x) := q(x)C_0(u, x),$$

$$C_0'(u, x) := \frac{\partial C_0(u, x)}{\partial u} = \left\{C_0^{kr}(u, x) := \frac{\partial C_0^k(u, x)}{\partial u_r}; k, r = \overline{1, d}\right\},$$

and the diffusion matrix

$$B(u) = B_0(u) - B_1(u), \qquad [4.149]$$

$$B_0(u) := 2\int_X \pi(dx)\overline{C}_0(u, x)R_0\overline{C}_0(u, x),$$

$$B_1(u) := q\int_X \rho(dx)C_0^T(u, x)C_0(u, x),$$

and w_t is a standard Wiener process.

THEOREM 4.4.– *We assume that the Lyapunov function $V(u), u \in R^d$, of the averaging system [4.144] satisfies the following conditions:*

$C1$: *the averaged system [4.145] is exponentially stable*

$$C(u)V'(u) \leq -c_0 V(u), c_0 > 0;$$

$C2$: *for* $\overline{C}(u) := \max_{x \in X}(|C(u,x)| + |C_0(u,x)|)$ *and* $w_k(u) := \overline{C}(u)w'_{k-1}, k = \overline{1,5}$, *boundedness*

$$|w_k(u)| \leq c_k(1 + V(u)), c_k > 0, k = \overline{1,5}.$$

holds true.

Finally, the normalizing function $a(t)$ is monotonically decreasing, bounded and satisfies the following conditions:

$$C3 : \int_0^\infty a(t)dt = \infty, \int_0^\infty a^2(t)dt < \infty.$$

Then, at each positive $\varepsilon \leq \varepsilon_0$, ε_0 is small enough, SAP [4.140] converges with probability 1 to a unique equilibrium point of the averaged evolutionary system [4.140]:

$$P\{\lim_{t \to \infty} u^\varepsilon(t) = 0\} = 1.$$

For jumping SAP [4.140], embeddings [4.142] form the extended Markov renewal process. The compensating operator for the process [4.142] on the test functions $\varphi(u, x, t), u \in R^d, x \in X, t \geq 0$, is defined by the conditional expectation

$$\boldsymbol{L}_t^\varepsilon \varphi(u, x, t) := \varepsilon^{-1} q(x) E[\varphi(u_{n+1}^\varepsilon, x_{n+1}^\varepsilon, \tau_{n+1}^\varepsilon) -$$

$$-\varphi(u, x, t) \,|u_n^\varepsilon = u,\ x_n^\varepsilon = x, \tau_n^\varepsilon = t]. \quad [4.150]$$

Furthermore, we prove two general statements.

LEMMA 4.13.– *The compensating operator [4.150] for the extended Markov renewal process [4.142] has the analytical representation*

$$\boldsymbol{L}_t^\varepsilon \varphi(u, x) := \varepsilon^{-2} Q\varphi(u, x) + \varepsilon^{-2} q(x) \boldsymbol{P}\boldsymbol{D}_t^\varepsilon(x)\varphi(u, x), \quad [4.151]$$

where the shift operator $\boldsymbol{D}_t^\varepsilon(x)$ has the form

$$\boldsymbol{D}_t^\varepsilon(x)\varphi(u, x) := \varphi(u + \varepsilon^2 a(t) C^\varepsilon(u, x), x) - \varphi(u, x), \quad [4.152]$$

and the operator $\boldsymbol{P}$ acts only at the second variable.

PROOF.– First, in accordance with [4.150], we calculate the conditional expectation

$$E[\varphi(u_{n+1}^\varepsilon, x_{n+1}^\varepsilon)|u_n^\varepsilon = u, x_n^\varepsilon = x, \tau_n^\varepsilon = t] = E_{u,x,t}\varphi(u_{n+1}^\varepsilon, x_{n+1}^\varepsilon)$$
$$= E_{u,x,t}\varphi(u_n^\varepsilon + \Delta u_{n+1}^\varepsilon, x_{n+1}^\varepsilon), \qquad [4.153]$$

where $\Delta u_{n+1}^\varepsilon := u_{n+1}^\varepsilon - u_n^\varepsilon$.

For calculating increment $\Delta u_{n+1}^\varepsilon$, we use the following representations from [4.140] and [4.142], i.e.

$$u_{n+1}^\varepsilon := u^\varepsilon(\tau_{n+1}^\varepsilon) = u + \varepsilon^2 \sum_{k=0}^{n} a_k^\varepsilon C^\varepsilon(u_k^\varepsilon, x_k^\varepsilon),$$

$$u_n^\varepsilon := u^\varepsilon(\tau_n^\varepsilon) = u + \varepsilon^2 \sum_{k=0}^{n-1} a_k^\varepsilon C^\varepsilon(u_k^\varepsilon, x_k^\varepsilon).$$

Considering the last relation, we have

$$\Delta u_{n+1}^\varepsilon = \varepsilon^2 a_n^\varepsilon C^\varepsilon(u_n^\varepsilon, x_n^\varepsilon). \qquad [4.154]$$

Substituting [4.154] into [4.153], we obtain

$$E_{u,x,t}\varphi(u_n^\varepsilon + \Delta u_{n+1}^\varepsilon, x_{n+1}^\varepsilon) = E_{u,x,t}\varphi(u + \varepsilon^2 a(t) C^\varepsilon(u,x), y),$$

where $y = x_{n+1}^\varepsilon$ is the value of the process $x(t), t \geq 0$, at the moment of next jump.

Considering the last equality, the compensating operator [4.150] has the form

$$\boldsymbol{L}_t^\varepsilon \varphi(u,x) := \varepsilon^{-2} q(x) E_{u,x,t}[\varphi(u + \varepsilon^2 a(t) C^\varepsilon(u,x), y) - \varphi(u,x)]$$
$$= \varepsilon^{-2} q(x) \int_X P(x,dy)[\varphi(u + \varepsilon^2 a(t) C^\varepsilon(u,x), y) - \varphi(u,x)].$$

Using summand $\pm\varphi(u,y)$, from the last equality, we have

$$\boldsymbol{L}_t^\varepsilon \varphi(u,x) = \varepsilon^{-2} q(x) \int_X P(x,dy)[\varphi(u,y) - \varphi(u,x)]$$
$$+ \varepsilon^{-2} q(x) \int_X P(x,dy)[\varphi(u + \varepsilon^2 a(t) C^\varepsilon(u,x), y) - \varphi(u,y)]. \qquad [4.155]$$

Considering [4.152] from [4.155], we obtain [4.151].

LEMMA 4.14.– *The compensating operator [4.151] of the embedded Markov renewal process [4.152] on the test functions* $\varphi(u,\cdot) \in C^3(R^d)$ *has the asymptotic representation*

$$\boldsymbol{L}_t^\varepsilon \varphi(u,x) = \varepsilon^{-2} Q\varphi(u,x) + \varepsilon^{-1} a(t) Q_1(x)\varphi(u,x) \\ + a(t)Q_2(x)\varphi(u,x) + \theta_L^\varepsilon(x)\varphi(u,x), \quad [4.156]$$

where

$$Q_1(x)\varphi(u,x) = \boldsymbol{C}_0(x)Q_0\varphi(u,x), \quad [4.157]$$

$$Q_2(x)\varphi(u,x) = \boldsymbol{C}(x)Q_0\varphi(u,x) + \frac{a(t)}{2}\boldsymbol{C}_0^2(x)Q_0\varphi(u,x), \quad [4.158]$$

$$\boldsymbol{C}_0(x)\varphi(u,x) := C_0(u,x)\varphi'_u(u,x), \boldsymbol{C}(x)\varphi(u,x) := C(u,x)\varphi'_u(u,x), \quad [4.159]$$

and the operator Q_0 *has the representation*

$$Q_0\varphi(x) = q(x)\boldsymbol{P}\varphi(x) = q(x)\int_X P(x,dy)\varphi(y). \quad [4.160]$$

The remainder term $\theta_L^\varepsilon(x)\varphi(u,x)$ *has the form*

$$\| \theta_L^\varepsilon(x)\varphi(u,x) \| \to 0, \varepsilon \to 0. \quad [4.161]$$

PROOF.– According to smoothness of functions $\varphi(u,x)$, we have decomposition

$$\boldsymbol{D}_t^\varepsilon(x)\varphi(u,x) = \varepsilon^2 a(t) C^\varepsilon(u,x)\varphi'_u(u,x) \\ + \varepsilon^4 \frac{a^2(t)}{2}(C^\varepsilon(u,x))^2\varphi''_u(u,x) + \varepsilon^2\theta_D^\varepsilon(x)\varphi(u,x) \quad [4.162]$$

for the operator $\boldsymbol{D}_t^\varepsilon(x)$, where

$$\theta_D^\varepsilon(x)\varphi(u,x) = \varepsilon^4 a^3(t)\frac{1}{6}(C^\varepsilon(u,x))^3\varphi'''_u(\theta u,x), 0 \le \theta \le 1. \quad [4.163]$$

Considering [4.144] from the last representation, we have

$$\boldsymbol{D}_t^\varepsilon(x)\varphi(u,x) = [\varepsilon a(t)C_0(u,x) + \varepsilon^2 a(t)C(u,x)]\varphi'_u(u,x) \\ + \varepsilon^2\frac{a^2(t)}{2}(C_0(u,x))^2\varphi''_u(u,x) + \varepsilon^4\frac{a^2(t)}{2}(C(u,x))^2\varphi''_u(u,x) \\ + \varepsilon^2\theta_D^\varepsilon(x)\varphi(u,x). \quad [4.164]$$

Substituting [4.164] into [4.151], we obtain [4.156], where the remainder term has the form

$$\theta_L^\varepsilon(x)\varphi(u,x) = \varepsilon^{-2}q(x)\boldsymbol{P}[\varepsilon^4\frac{a^2(t)}{2}(C(u,x))^2\varphi''_u(u,x) + \varepsilon^2\theta_D^\varepsilon(x)\varphi(u,x)],$$

that is, in accordance with [4.163] and [4.160],

$$\theta_L^\varepsilon(x)\varphi(u,x) = \varepsilon^2 \frac{a^2(t)}{2}(C(u,x))^2 Q_0 \varphi_u''(u,x)$$
$$+ \varepsilon^4 a^3(t)\frac{1}{6}(C^\varepsilon(u,x))^3 Q_0 \varphi_u'''(\theta u,x). \qquad [4.165]$$

According to [4.144] and the boundedness of multipliers on the right-hand side of [4.165], except ε, we obtain the property [4.161] and the lemma is proved.

PROOF.– We prove the theorem based on the asymptotic representation for the compensating operator [4.156] using the perturbed Lyapunov function

$$V^\varepsilon(u,x) = V(u) + \varepsilon a(t) V_1(u,x) + \varepsilon^2 a(t) V_2(u,x). \qquad [4.166]$$

LEMMA 4.15.– *The compensating operator [4.156] on functions [4.166] has the asymptotic representation*

$$\boldsymbol{L}_t^\varepsilon V^\varepsilon(u,x) = a(t)C(u)V'(u) + a^2(t)\boldsymbol{L}_0 V(u) + \varepsilon a^2(t) H_L^\varepsilon(x,t)V(u), \quad [4.167]$$

where

$$\boldsymbol{L}_0 V(u) = b(u)V'(u) + \frac{1}{2} Sp[B(u)V''(u)]$$

and the remainder term $H_L^\varepsilon(x,t)V(u)$ is bounded.

PROOF.– For the operator [4.156], we observe

$$\boldsymbol{L}_{t0}^\varepsilon \varphi(u,x) = \varepsilon^{-2} Q\varphi(u,x) + \varepsilon^{-1} a(t) Q_1(x)\varphi(u,x) + a(t) Q_2(x)\varphi(u,x). \qquad [4.168]$$

The operator $\boldsymbol{L}_{t0}^\varepsilon$ on functions [4.166] can be represented in the form

$$\boldsymbol{L}_{t0}^\varepsilon V^\varepsilon(u,x) = [\varepsilon^{-2} Q + \varepsilon^{-1} a(t) Q_1(x) + a(t) Q_2(x)]$$
$$= [V(u) + \varepsilon a(t) V_1(u,x) + \varepsilon^2 a(t) V_2(u,x)]$$
$$= \varepsilon^{-2} Q V(u) + \varepsilon^{-1} a(t)[Q V_1(u,x) + Q_1(x) V(u)] + Q V_2(u,x)$$
$$+ a(t) Q_1(x) V_1(u,x) + a(t) Q_2 V(u)(x) + \varepsilon \theta_0^\varepsilon(x) V(u), \qquad [4.169]$$

where

$$\theta_0^\varepsilon(x)V(u) = a(t) Q_1(x) V_2(u,x) + a(t) Q_2(x) V_1(u,x)$$
$$+ \varepsilon a(t) Q_2(x) V_2(u,x). \qquad [4.170]$$

From the solvability condition of decomposition [4.169], we obtain the equations

$$QV(u) = 0, \qquad [4.171]$$

$$QV_1(u,x) + Q_1(x)V(u) = 0, \qquad [4.172]$$

$$QV_2(u,x) + a(t)Q_1(x)V_1(u,x) + a(t)Q_2(x)V(u) = \boldsymbol{L}_t V(u). \qquad [4.173]$$

Equation [4.170] holds true, since the function $V(u)$ does not depend on x, i.e. the function $V(u)$ belongs to the space of zeros N_Q of the operator Q.

The balance condition BC1 defines the solvability of equation [4.172], since

$$\Pi Q_1(x)V(u) = \Pi \boldsymbol{C}_0(x)Q_0V(u) = \Pi C_0(u,x)q(x)\boldsymbol{P}V'(u)$$
$$= V'(u)\int_X \pi(dx)q(x)C_0(u,x) = qV'(u)\int_X \rho(dx)C_0(u,x) = 0.$$

Therefore, the solution of equation [4.172] has the form

$$V_1(u,x) = R_0\overline{\boldsymbol{C}}_0(x)V(u), \qquad [4.174]$$

where $\overline{\boldsymbol{C}}_0(x) := q(x)\boldsymbol{C}_0(x)$.

Furthermore, we substitute [4.174] into [4.173] and obtain

$$QV_2(u,x) + \boldsymbol{L}_t(x)V(u) = \boldsymbol{L}_t V(u), \qquad [4.175]$$

where

$$\boldsymbol{L}_t(x)V(u) = a(t)Q_1(x)V_1(u,x) + a(t)Q_2(x)V(u). \qquad [4.176]$$

We calculate the right-hand side of [4.176]. Using [4.174] for the first summand, we have

$$Q_1(x)V_1(u,x) = \boldsymbol{C}_0(x)Q_0a(t)R_0\overline{\boldsymbol{C}}_0(x)V(u)$$
$$= a(t)\boldsymbol{C}_0(x)q(x)\boldsymbol{P}R_0\overline{\boldsymbol{C}}_0(x)V(u). \qquad [4.177]$$

In [4.177], we use the representation

$$\boldsymbol{P}R_0 = R_0 + g(x)[\Pi - I].$$

Considering $\overline{\boldsymbol{C}}_0(x) = q(x)\boldsymbol{C}_0(x)$, we obtain

$$\boldsymbol{P}R_0\overline{\boldsymbol{C}}_0(x)V(u) = [R_0 + g(x)[\Pi - I]]\overline{\boldsymbol{C}}_0(x)V(u)$$
$$= [R_0\overline{\boldsymbol{C}}_0(x) + g(x)\Pi\,\overline{\boldsymbol{C}}_0(x) - g(x)\overline{\boldsymbol{C}}_0(x)]V(u).$$

Since

$$\Pi\overline{\boldsymbol{C}}_0(x) = \Pi q(x)\boldsymbol{C}_0(x) = \int\limits_X C_0(u,x)\pi(dx)q(x) = q\int\limits_X \rho(dx)C_0(u,x) = 0,$$

for [4.187], we obtain

$$Q_1(x)V_1(u,x) = a(t)[\overline{\boldsymbol{C}}_0(x)R_0\overline{\boldsymbol{C}}_0(x)V(u) - \boldsymbol{C}_0(x)\overline{\boldsymbol{C}}_0(x)V(u)]. \qquad [4.178]$$

We calculate the second summand in [4.176]:

$$\begin{aligned} Q_2(x)V(u) &= \boldsymbol{C}(x)Q_0V(u) + \frac{a(t)}{2}\boldsymbol{C}_0^2(x)Q_0V(u) \\ &= \overline{\boldsymbol{C}}(x)V(u) + \frac{a(t)}{2}\boldsymbol{C}_0^2(x)q(x)V(u), \end{aligned}$$

i.e.

$$Q_2(x)V(u) = \overline{\boldsymbol{C}}(x)V(u) + \frac{a(t)}{2}\boldsymbol{C}_0(x)\overline{\boldsymbol{C}}_0(x)V(u). \qquad [4.179]$$

Consequently, [4.178] and [4.179] for [4.176] give the representation

$$\begin{aligned} \boldsymbol{L}_t(x)V(u) &= a^2(t)\overline{\boldsymbol{C}}_0(x)R_0\overline{\boldsymbol{C}}_0(x)V(u) - a^2(t)\boldsymbol{C}_0(x)\overline{\boldsymbol{C}}_0(x)V(u) \\ &\quad + a(t)\,\overline{\boldsymbol{C}}(x)V(u) + \frac{a^2(t)}{2}\boldsymbol{C}_0(x)\overline{\boldsymbol{C}}_0(x)V(u) \\ &= a(t)\overline{\boldsymbol{C}}(x)V(u) + a^2(t)\boldsymbol{L}_0(x)V(u), \end{aligned} \qquad [4.180]$$

where

$$\boldsymbol{L}_0(x) := \overline{\boldsymbol{C}}_0(x)R_0\overline{\boldsymbol{C}}_0(x) - \frac{1}{2}\boldsymbol{C}_0(x)\overline{\boldsymbol{C}}_0(x). \qquad [4.181]$$

The limit operator $\boldsymbol{L}_t$ for the operator [4.180] has the form $\boldsymbol{L}_t = \Pi\boldsymbol{L}_t(x)\Pi$. Therefore, we have

$$\boldsymbol{L}_t := a(t)\Pi\overline{\boldsymbol{C}}(x)\Pi + a^2(t)\Pi\boldsymbol{L}_0(x)\Pi. \qquad [4.182]$$

We calculate the right-hand side of the representation [4.182]. For the first summand, we calculate for the multiplier

$$\begin{aligned} \Pi\overline{\boldsymbol{C}}(x)\Pi V(u) &= \int\limits_X \pi(dx)C(u,x)q(x)V'(u) \\ &= q\int\limits_X \rho(dx)C(u,x)V'(u) = \tilde{\boldsymbol{N}}V(u), \end{aligned} \qquad [4.183]$$

where $\boldsymbol{C}V(u) = C(u)V'(u)$.

Using [4.181] for the second summand, we consider the multiplier

$$\Pi\boldsymbol{L}_0(x)\Pi V(u) = \Pi\overline{\boldsymbol{C}}_0(x)R_0\overline{\boldsymbol{C}}_0(x)V(u) - \frac{1}{2}\Pi\boldsymbol{C}_0(x)\overline{\boldsymbol{C}}_0(x)V(u)$$

$$= \Pi\overline{\boldsymbol{C}}_0(x)R_0C_0(u,x)q(x)V'(u) - \frac{1}{2}\Pi\boldsymbol{C}_0(x)C_0(u,x)q(x)V'(u)$$

$$= \Pi C_0(u,x)q(x)R_0[C_0(u,x)q(x)V'(u)]' - \frac{1}{2}\Pi C_0(u,x)[C_0(u,x)q(x)V'(u)]'$$

$$= \Pi C_0(u,x)q(x)R_0q(x)[C_0'(u,x)V'(u) + C_0(u,x)V''(u)]$$

$$- \frac{1}{2}\Pi C_0(u,x)q(x)[C_0'(u,x)V'(u) + C_0(u,x)V''(u)]$$

$$= [\Pi C_0(u,x)q(x)R_0q(x)C_0'(u,x) - \frac{1}{2}\Pi C_0(u,x)q(x)C_0'(u,x)]V'(u)$$

$$+ [\Pi C_0(u,x)q(x)R_0q(x)C_0(u,x) - \frac{1}{2}\Pi C_0(u,x)q(x)C_0(u,x)]V''(u).$$

Furthermore,

$$\Pi\boldsymbol{L}_0(x)\Pi V(u) = b(u)V'(u) + \frac{1}{2}Sp[B(u)V''(u)],$$

where $b(u)$ and $B(u)$ are represented in [4.148] and [4.149].

Using the last equality and the relation [4.183] for the right-hand side of [4.182], we have the representation

$$\boldsymbol{L}_t = a(t)\boldsymbol{C}V(u) + a^2(t)\boldsymbol{L}_0V(u), \qquad [4.184]$$

where

$$\boldsymbol{L}_0V(u) = b(u)V'(u) + \frac{1}{2}Sp[B(u)V''(u)]. \qquad [4.185]$$

Consequently, from [4.169], we obtain

$$\boldsymbol{L}_{t0}^{\varepsilon}V^{\varepsilon}(u,x) = a(t)\boldsymbol{C}V(u) + a^2(t)\boldsymbol{L}_0V(u) + \varepsilon\theta_0^{\varepsilon}(x)V(u). \qquad [4.186]$$

Finally, to complete the analysis of the remainder term in [4.169], we use the representation

$$V_2(u,x) = \boldsymbol{L}_{ta}(x)V(u), \qquad [4.187]$$

where

$$\boldsymbol{L}_{ta}(x)V(u) = \widetilde{\boldsymbol{C}}(x) + a(t)\widetilde{\boldsymbol{L}}_0(x)$$

and

$$\widetilde{\boldsymbol{C}}(x) = \overline{\boldsymbol{C}}(x) - \boldsymbol{C}, \widetilde{\boldsymbol{L}}_0(x) = \boldsymbol{L}_0(x) - \boldsymbol{L}_0.$$

Using [4.187] and [4.174] in [4.170], we obtain the boundedness of the multiplier $\theta_0^\varepsilon(x)V(u)$, whence

$$\varepsilon\|\theta_0^\varepsilon(x)V(u)\| \to 0, \varepsilon \to 0. \qquad [4.188]$$

For further analysis, we need to have a clear form for $\theta_0^\varepsilon(x)V(u)$. In [4.170], we calculate the first summand:

$$\begin{aligned} Q_2(x)V_1(u,x) &= [a(t)\boldsymbol{C}(x)Q_0 + \frac{a^2(t)}{2}\boldsymbol{C}_0^2(x)Q_0]a(t)R_0\overline{\boldsymbol{C}}_0(x)V(u) \\ &= a^2(t)\theta_1(x)V(u), \end{aligned} \qquad [4.189]$$

where

$$\theta_1(x)V(u) = \theta_{11}(x)V(u) + \frac{a(t)}{2}\theta_{12}(x)V(u), \qquad [4.190]$$

and

$$\theta_{11}(x) = \overline{\boldsymbol{C}}(x)R_0\overline{\boldsymbol{C}}_0(x) - \boldsymbol{C}(x)\overline{\boldsymbol{C}}_0(x),$$

$$\theta_{12}(x) = \boldsymbol{C}_0(x)\overline{\boldsymbol{C}}_0(x)R_0\overline{\boldsymbol{C}}_0(x) - \boldsymbol{C}(x)\overline{\boldsymbol{C}}_0^2(x).$$

For the second summand:

$$\begin{aligned} Q_2(x)V_2(u,x) &= [a(t)\boldsymbol{C}(x)Q_0 + \frac{a^2(t)}{2}\boldsymbol{C}_0^2(x)Q_0]a(t)\boldsymbol{L}_{ta}(x)V(u) \\ &= a^2(t)[\boldsymbol{C}(x)Q_0\boldsymbol{L}_{ta}(x) + \frac{a(t)}{2}\boldsymbol{C}_0^2(x)Q_0\boldsymbol{L}_{ta}(x)]V(u). \end{aligned} \qquad [4.191]$$

We calculate summands in [4.191]. For the first one, we obtain

$$\tilde{\boldsymbol{N}}(x)Q_0\boldsymbol{L}_{ta}(x) = \boldsymbol{C}(x)q(x)\boldsymbol{P}\boldsymbol{L}_{ta}(x). \qquad [4.192]$$

Since

$$\boldsymbol{P}\boldsymbol{L}_{ta}(x) = \boldsymbol{P}[\widetilde{\boldsymbol{C}}(x) + a(t)\widetilde{\boldsymbol{L}}_0] = a(t)\boldsymbol{L}_{0P}(x), \qquad [4.193]$$

where

$$\boldsymbol{L}_{0P}(x) := \boldsymbol{P}\boldsymbol{L}_0(x) - \boldsymbol{L}_0,$$

then from [4.192], we have

$$\tilde{\boldsymbol{N}}(x)Q_0\boldsymbol{L}_{ta}(x) = a(t)\overline{\boldsymbol{C}}(x)\boldsymbol{L}_{0P}(x). \quad [4.194]$$

Using [4.194] for the second summand, we have

$$\begin{aligned}\frac{a(t)}{2}\boldsymbol{C}_0^2(x)Q_0\boldsymbol{L}_{ta}(x) &= \frac{a(t)}{2}\boldsymbol{C}_0^2(x)q(x)a(t)\boldsymbol{L}_{0P}(x)\\ &= \frac{a^2(t)}{2}\boldsymbol{C}_0(x)\overline{\boldsymbol{C}}_0(x)\boldsymbol{L}_{0P}(x). \quad [4.195]\end{aligned}$$

From [4.194] and [4.195], we have the representation for [4.191]:

$$\begin{aligned}Q_2(x)V_2(u,x) &= a^2(t)[a(t)\overline{\boldsymbol{C}}(x)\boldsymbol{L}_{0P}(x) + \frac{a^2(t)}{2}\boldsymbol{C}_0(x)\overline{\boldsymbol{C}}_0(x)\boldsymbol{L}_{0P}(x)]V(u)\\ &= a^3(t)\theta_2(x)V(u), \quad [4.196]\end{aligned}$$

where

$$\theta_2(x) := \overline{\boldsymbol{C}}(x)\boldsymbol{L}_{0P}(x) + \frac{a(t)}{2}\boldsymbol{C}_0(x)\overline{\boldsymbol{C}}_0(x)\boldsymbol{L}_{0P}(x). \quad [4.197]$$

The third summand in [4.170] has the form

$$Q_1(x)V_2(u,x) = a(t)\boldsymbol{C}_0(x)Q_0a(t)\,\boldsymbol{L}_{ta}(x)V(u) = a^3(t)\overline{\boldsymbol{C}}_0(x)\,\boldsymbol{L}_{0P}(x). \quad [4.198]$$

For [4.170], the relations [4.189], [4.196] and [4.198] give

$$\theta_0^\varepsilon(x)V(u) = a^2(t)\theta_1^\varepsilon(x)V(u), \quad [4.199]$$

where

$$\theta_1^\varepsilon(x)V(u) = [\theta_1(x) + a(t)\overline{\boldsymbol{C}}_0(x)\boldsymbol{L}_{0P}(x) + \varepsilon a(t)\theta_2(x)]V(u), \quad [4.200]$$

where $\theta_1(x)$ and $\theta_2(x)$ are defined in [4.190] and [4.197].

We consider the full operator [4.156] using a truncated one [4.168] in the form

$$\boldsymbol{L}_t^\varepsilon = \boldsymbol{L}_{t0}^\varepsilon + \theta_{\boldsymbol{L}}^\varepsilon(x)$$

on the test functions $V^\varepsilon(u,x)$. We obtain the following decomposition:

$$\begin{aligned}&\boldsymbol{L}_t^\varepsilon V^\varepsilon(u,x) = [\,\boldsymbol{L}_{t0}^\varepsilon + \theta_{\boldsymbol{L}}^\varepsilon(x)]V^\varepsilon(u,x) = \boldsymbol{L}_{t0}^\varepsilon V^\varepsilon(u,x) + \theta_{\boldsymbol{L}}^\varepsilon(x)V^\varepsilon(u,x)\\ &= a(t)\boldsymbol{C}V(u) + a^2(t)\boldsymbol{L}_0V(u) + \varepsilon\theta_0^\varepsilon(x)V(u) + \theta_L^\varepsilon(x)\Phi^\varepsilon(x)V(u),\\ &\Phi^\varepsilon(x)V(u) = I + \varepsilon a(t)R_0\overline{\boldsymbol{C}}_0(x) + \varepsilon^2a(t)\boldsymbol{L}_{ta}(x).\end{aligned}$$

Considering the last relation, we obtain the limit representation for the compensating operator

$$\boldsymbol{L}_t^\varepsilon V^\varepsilon(u,x) = a(t)\boldsymbol{C}V(u) + a^2(t)\boldsymbol{L}_0 V(u) + \varepsilon a^2(t) H_t^\varepsilon(u,x)V(u), \qquad [4.201]$$

where

$$H_t^\varepsilon(u,x)V(u) = \theta_1^\varepsilon(x)V(u) + \theta_2^\varepsilon(x)V(u), \qquad [4.202]$$

and

$$\theta_1^\varepsilon(x) = \theta_1(x) + a(t)\overline{\boldsymbol{C}}_0(x)\boldsymbol{L}_{0t}(x) + \varepsilon\theta_2(x),$$

$$\theta_2^\varepsilon(x)V(u) = \frac{\varepsilon}{2}C^2(u,x)q(x)\Phi^\varepsilon(x)V''(u)$$
$$+\frac{a(t)}{6}(C_\varepsilon(u,x))^3 q(x)\Phi^\varepsilon(x)V'''(\theta u), 0 \le \theta \le 1,$$

where $C_\varepsilon(u,x) = \varepsilon C^\varepsilon(u,x) = \varepsilon C(u,x) + C_0(u,x)$.

The smoothness of the function $V(u), C(u,x), C_0(u,x)$, as well as the boundedness $a(t) \le a < \infty$ of the function $a(t)$ and the operators $\boldsymbol{P}$ and R_0 guarantee the boundedness of the right-hand side of [4.202].

PROOF OF THEOREM 4.3.– Using estimations $C2$ and condition $C1$ of exponential stability for the system [4.144], and monotonicity and boundedness of $a(t) \le a < \infty$ for normalizing functions, we have

$$\boldsymbol{L}_t^\varepsilon V^\varepsilon(u,x) = a(t)\boldsymbol{C}V(u) + a^2(t)\boldsymbol{L}_0 V(u) + \varepsilon a^2(t) H_t^\varepsilon(u,x)V(u)$$
$$\le -a(t)(\delta^* - \varepsilon c)V(u) \le -\delta a(t)V(u), \delta > 0. \qquad [4.203]$$

The last inequality holds true for each $\varepsilon \le \varepsilon_0$, where for ε_0, the inequality $\varepsilon_0 < \delta^*/c$ holds true.

Second, the clear form of the perturbation [4.174] and [4.187], as well as condition $C2$ of the theorem, gives the estimation for the perturbed Lyapunov function

$$0 < (1 - \varepsilon a(t)c)V(u) \le V^\varepsilon(u,x) \le (1 + \varepsilon a(t)c)V(u).$$

Third, in accordance with Lemma 4.14, it follows that the process

$$V_n^\varepsilon := V^\varepsilon(u_n^\varepsilon, x_n^\varepsilon), n \ge 0$$

is a non-negative supermartingale.

Using estimates [4.203], the proof of the theorem is completed by the scheme of the proof of Theorem 4.1.

4.3. Asymptotic normality of fluctuations of the procedure of stochastic approximation with diffusive perturbation in a Markov environment

We consider the asymptotic normality of a continuous procedure of stochastic approximation in the case where the regression function contains a singularly perturbed term depending on the external medium described by a uniformly ergodic Markov process. Within the framework of the scheme of diffusion approximation, we formulate sufficient conditions for asymptotic normality in terms of the existence of a Lyapunov function for the corresponding averaged equation. Introduction procedures of stochastic approximation are widely used in the solution of problems of optimization in mathematical statistics, control theory, communication theory, the theory of pattern recognition and so on (Nevelson and Khasminsky 1972). Along with the main problem of the convergence of a procedure of stochastic approximation to the root of the regression equation, there is an important problem of the estimation of the rate of convergence, which follows from the asymptotic normality of a fluctuation of the procedure of stochastic approximation in the neighborhood of the root of the regression equation. Investigations of the asymptotic normality of fluctuations in the classical schemes of procedures of stochastic approximation (Nevelson and Khasminsky 1972; Korolyuk and Korolyuk 1999) are carried out by using the principle of invariance for sums (for a discrete procedure of stochastic approximation) and processes (for a continuous procedure of stochastic approximation). In Chabanyuk (2004a), the asymptotic normality of a procedure of stochastic approximation in a Markov medium in the scheme of averaging was studied by using the martingale characterization of the corresponding two-component Markov process and the solution of the singular perturbation problem for the generator of this process. As a result, the generator of a limit diffusion process of the Ornstein–Uhlenbeck type was obtained. In this chapter, we consider the asymptotic normality of a continuous procedure of stochastic approximation in a Markov medium with asymptotically diffusive perturbations under the condition of the convergence of this procedure (Korolyuk and Limnios 2005). We use the second method of Lyapunov functions for the averaged system.

1. Statement of the problem. A continuous procedure of stochastic approximation with asymptotically diffusive perturbation in the scheme of series with a small parameter of the series e > 0 is defined by the evolution equation

$$\frac{du^\varepsilon(t)}{dt} = a(t)\left[C(u^\varepsilon(t), x(\frac{t}{\varepsilon^4})) + \varepsilon^{-1}C_0(x(\frac{t}{\varepsilon^4}))\right] \qquad [4.204]$$

on the real axis $R = (-\infty; +\infty)$. The regression function $C(u, x)$, $u \in \mathbb{R}$, $x \in X$, and the perturbation function $C_0(x)$, $x \in X$, satisfy the conditions for the existing global solution for the accompanying systems

$$\frac{du_x^\varepsilon(t)}{dt} = C^\varepsilon(u_x^\varepsilon(t), x), \ x \in X,$$
$$C^\varepsilon(u, x) = C(u, x) + \varepsilon^{-1}C_0(x).$$

A Markov process $x(t)$, $t \geq 0$, in the standard space $(X, \mathbf{X})$ with stationary distribution $\pi(B)$, $B \in \mathbf{X}$, is given by the generator $\mathbf{Q}$ defined by the relation

$$\mathbf{Q}\varphi(x) = \int_X P(x, dy)[\varphi(y) - \varphi(x)],$$

where $\varphi(x) \in D_Q$ is the domain of definition of the operator $\mathbf{Q}$. Moreover, the generator $\mathbf{Q}$ is a reducibly invertible operator for which there exists a potential R_0 defined by the equation

$$R_0 Q = Q R_0 = \Pi - I,$$

where Π is the projector in the Banach space $\mathbf{B}(\mathbf{X})$ of real-valued functions with the supremum norm defined by the relation

$$\Pi\varphi(x) := \tilde{\varphi}\mathbf{1}(x), \ \tilde{\varphi} := \int_X \pi(dx)\varphi(x), \ \mathbf{1}(x) \equiv 1, \ x \in X.$$

The stationary distribution $\pi(B)$, $B \in \mathbf{X}$, is defined by the equation

$$\pi(B) = \int_X \pi(dy)Q(y, B), \ B \in \mathbf{X}, \ \pi(X) = 1.$$

If the function $a(t)$, $t \geq 0$, satisfies the corresponding conditions and the Markov process $x(t)$, $t \geq 0$, with the stationary distribution $\pi(B)$, $B \in \mathbf{X}$, is uniformly ergodic, then the procedure of stochastic approximation [4.204] converges with probability 1 to the equilibrium point u_0 of the averaged system

$$\frac{du(t)}{dt} = C(u(t)),$$

where $C(u)$ is the regression function $C(u, x)$ averaged over the stationary distribution $\pi(B)$, $B \in \mathbf{X}$:

$$C(u) := \int_X \pi(dx)C(u, x).$$

In the following, without loss of generality, we assume that $u_0 = 0$, i.e.

$$C(0) = 0. \qquad [4.205]$$

The perturbation $C_0(x)$ of the regression function $C(u, x)$ satisfies the following balance condition:

(BC1) $\int_X \pi(dx)C_0(x) = 0$ or $\Pi C_0(x) = 0$.

We consider the standard procedure of stochastic approximation [4.204] with the normalizing function $a(t) = \frac{a}{t}$, $a > 0, t > 0$, and use the regression function $C(u, x)$ such that

(C1) $C(u, \cdot) \in C^2(\mathbb{R})$

and the second derivative with respect to u satisfies the global Lipschitz condition

(C2) $|C''_u(u, \cdot) - C''_u(u', \cdot)| \leq C|u - u'|$,

where the constant C does not depend on $x \in X$.

Then, using the Taylor formula, we obtain the following expansion for the regression function $C(u, x)$:

$$C(u, x) = C^0(x) + uC^1(x) + u^2 C_2(u, x), \tag{4.206}$$

where

$$\begin{aligned} &C^0(x) := C(0, x), \; C^1 := C'_u(0, x), \\ &C_2(u, x) := \tfrac{1}{2} C''_u(\theta u, x), \; 0 \leq \theta \leq 1. \end{aligned} \tag{4.207}$$

For $C^0(x)$, using [4.205] and the projector Π, we obtain one more balance condition:

(BC2) $\Pi C^0(x) = 0$.

Assume that the following condition for the convergence of the procedure of stochastic approximation [4.204] in a Markov medium is satisfied: there exists a Lyapunov function $V(u)$, $u \in \mathbb{R}$, that guarantees the exponential stability of system (2):

(S1) $C(u)V'(u) \leq -c_0 V(u)$, $c_0 > 0$,

and the function $\tilde{C}(u, x) = C(u, x) - C(u)$ satisfies the following additional conditions:

(S2) $|C_0(x) R_0 C_0(x) V''(u)| \leq c_1(1 + V(u))$, $c_1 > 0$,

$|C(u, x) R_0[\tilde{C}(u, x) V'(u)]'| \leq c_2 V(u)$, $c_2 > 0$,

$|C_0(x) R_0[\tilde{C}(u, x) V'(u)]'| \leq c_3(1 + V(u))$, $c_3 > 0$,

$|C_0(x) R_0 C_0(x) V'''(u)| \leq c_4(1 + V(u))$, $c_4 > 0$,

$|C(u, x) R_0 C_0(x) V'''(u)| \leq c_5 V(u)$, $c_5 > 0$.

NOTE 4.1.– *We can verify that, for the diffusive perturbation*

$$C_0^\varepsilon(t) = \varepsilon^{-2} a \int_{t_0}^{t} C_0(x(\frac{s}{\varepsilon^4}))\frac{ds}{s}, \quad [4.208]$$

the following weak convergence takes place:

$$C_0^\varepsilon(t) \Rightarrow \sigma(t) w(t), \ \varepsilon \to 0, \quad [4.209]$$

where

$$\sigma(t) = \frac{a\rho}{t}, \ \rho^2 = 2 \int_X \pi(dx) C_0(x) R_0 C_0(x),$$

and $w(t)$ is a standard Wiener process.

Denote

$$c := - \int_X \pi(dx) C^1(x), \ b := 1 + \frac{1}{2ac},$$

where $C^1(x)$ is defined in [4.207] and

$$C^1(v, w) := vb + \frac{\sqrt{t} w}{c}.$$

Consider a fluctuation of the procedure of stochastic approximation [4.204] in the form

$$v^\varepsilon(t) = \varepsilon^{-1} \sqrt{t}[u^\varepsilon(t) - \varepsilon C_0^\varepsilon(t)]. \quad [4.210]$$

NOTE 4.2.– *The reasons for centering and normalization in [4.210] follow from condition [4.205] and convergence [4.209].*

THEOREM 4.5.– *We suppose that conditions (S1) and (S2) are satisfied, the procedure of stochastic approximation [4.204] converges and the following additional conditions are satisfied:*

(D1) $\rho^2 > 0$;

(D2) $b < 0$.

Then, the following weak convergence takes place in every finite interval $(0 < t_0 < t < T)$*:*

$$v^\varepsilon(t) \Rightarrow \zeta(t), \ \varepsilon \to 0. \quad [4.211]$$

The limit process $\zeta(t)$, $t \geq 0$, is the Ornstein–Uhlenbeck diffusion process defined by the generator

$$\mathbf{L}\varphi(v,w) = \frac{a^2\rho^2}{2t^2}\varphi''_w(v,w) + \frac{1}{t}C^1(v,w)\varphi'_v(v,w). \qquad [4.212]$$

NOTE 4.3.– *The limit diffusion process $\zeta(t)$, $t \geq 0$, satisfies the stochastic equation*

$$d\zeta(t) = \frac{a}{t}[cb\zeta(t) + \rho\sqrt{t}w(t)]dt.$$

COROLLARY 4.2.– *Under the conditions of the theorem, the fluctuation $v^\varepsilon(t)$ has the asymptotically normal distribution $N(0,\sigma^2)$ for $a >> 0$, i.e.*

$$v^\varepsilon(t) \Rightarrow \zeta(t), \ t \to \infty.$$

and the random variable v belongs to $N(0,\sigma^2)$ with variance $\sigma^2 = a^2\rho^2/bc$.

Furthermore, we investigate the properties of the normalized procedure of stochastic approximation [4.210].

LEMMA 4.16.– *The normalized procedure of stochastic approximation [4.210] satisfies the evolution equation*

$$dv^\varepsilon(t) = \varepsilon^{-1}\frac{a}{\sqrt{t}}C(\varepsilon(\frac{v^\varepsilon(t)}{\sqrt{t}} + C_0^\varepsilon(t)), x(\frac{t}{\varepsilon^4}))dt + \frac{1}{2t}v^\varepsilon(t)dt. \qquad [4.213]$$

PROOF.– Consider the normalized fluctuation

$$\tilde{u}^\varepsilon(t) := \varepsilon^{-1}[u^\varepsilon(t) - \varepsilon C_0^\varepsilon(t)], \qquad [4.214]$$

which has the representation

$$d\tilde{u}^\varepsilon(t)\tilde{u}^\varepsilon(t) = \varepsilon^{-1}\frac{a}{t}C(u^\varepsilon(t), x(\frac{t}{\varepsilon^4}))dt. \qquad [4.215]$$

Using [4.214] and [4.210], we obtain $\tilde{u}^\varepsilon(t) = v^\varepsilon(t)/\sqrt{t}$. Taking this into account and calculating the differential on the left-hand side of [4.215], we obtain [4.213].

COROLLARY 4.3.– *For the evolution equation [4.213], the asymptotic expansion*

$$dv^\varepsilon(t) = (\varepsilon^{-1}\frac{a}{\sqrt{t}}C^0(x(\frac{t}{\varepsilon^4})) + \frac{a}{t}[b(x(\frac{t}{\varepsilon^4})v^\varepsilon(t) + \sqrt{t}C_0^\varepsilon(t)C^1(x)])dt$$
$$+ \theta_C((v^\varepsilon(t), x(\frac{t}{\varepsilon^4}))dt$$

with the negligible term

$$|\theta_C((v^\varepsilon(t), x)| \to 0, \ \varepsilon \to 0, \ t_0 < t < T,$$

is true. Here, $b(x) = C^1(x) + 1/2$.

PROOF.– On the right-hand side of [4.213], we use expansion [4.206] for the regression function $C(u, x)$.

Furthermore, we consider the generator of an extended Markov process

$$v^\varepsilon(t), C_0^\varepsilon(t), x_t^\varepsilon(\frac{t}{\varepsilon^4}),\ t \geq 0. \qquad [4.216]$$

LEMMA 4.17.– *The generator of process [4.216] on the test functions* $\varphi(v, w, \cdot) \in C^3(R \times R)$ *has the asymptotic representation*

$$\mathbf{L}_t^\varepsilon \varphi(v, w, x) = \varepsilon^{-4}\mathbf{Q}\varphi(\cdot, \cdot, x) + \varepsilon^{-2}\frac{a}{t}\mathbf{C}_0(x)\varphi(\cdot, w, \cdot) + \varepsilon^{-1}\frac{a}{\sqrt{t}}\mathbf{C}^0(x)\varphi(v, \cdot, \cdot)$$
$$+ \frac{a}{t}\mathbf{C}^1(v, w, x)\varphi(v, \cdot, \cdot) + \theta_t^\varepsilon(v, w, x)\varphi(v, w, x), \qquad [4.217]$$

where

$$\mathbf{C}_0(x)\varphi(w) = C_0(x)\varphi'(w), \qquad [4.218]$$

$$\mathbf{C}^0(x)\varphi(v) = C^0(x)\varphi'(v), \qquad [4.219]$$

$$\mathbf{C}^1(v, w, x)\varphi(v) = ((v + \sqrt{t}w)C^1(x) + \frac{v}{2a})\varphi'(v), \qquad [4.220]$$

and the remainder $\theta_t^\varepsilon(v, w, x)\varphi(v, w, x)$ *is such that*

$$\|\theta_t^\varepsilon(v, w, x)\varphi(v, w, x)\| \to 0,\ \varepsilon \to 0.$$

PROOF.– According to the definition of the generator of the extended Markov process [4.216], it is necessary to determine the conditional mathematical expectation. Using [4.208], we get

$$\mathbb{E}_{v,w,x}[\varphi(v + \Delta v^\varepsilon(t), w + \Delta C_0^\varepsilon(t), x_{t+\Delta}^\varepsilon(t)) - \varphi(v, w, x)]$$
$$= \mathbf{L}_t^\varepsilon \varphi(v, w, x)\Delta + o(\Delta),$$

where the generator $\mathbf{L}_t^\varepsilon$ has the form [4.217], $o(\Delta)$ is an infinitely small quantity of the second order with respect to Δ, and Δ in front of a function is understood as the operator of the increment of the function.

The next step of our investigation is to solve the singular perturbation problem. To construct the limit operator, we consider a solution of the singular perturbation problem (Korolyuk and Limnios 2005) for the operator $\mathbf{L}_t^\varepsilon$ in the form [4.217] on the perturbed test function

$$\varphi^\varepsilon(v, w, x) = \varphi(v, w) + \varepsilon^2\frac{1}{t}\varphi_2(v, w, x) + \varepsilon^3\frac{1}{\sqrt{t}}\varphi_3(v, w, x) + \varepsilon^4\frac{1}{t}\varphi_4(v, w, x).$$

LEMMA 4.18.– *A solution of the singular perturbation problem for the generator [4.217] has the form*

$$\mathbf{L}_t^\varepsilon \varphi^\varepsilon(v,w,x) = \frac{1}{t}\mathbf{L}\varphi(v,w) + \theta_L^\varepsilon(v,w,x)\varphi(v,w), \quad [4.221]$$

where the operator $\mathbf{L}$ *is defined by [4.212] and the remainder* $\theta_L^\varepsilon(v,w,x)\varphi(v,w)$ *is such that*

$$\|\theta_L^\varepsilon(v,w,x)\varphi(v,w)\| \to 0, \ \varepsilon \to 0.$$

PROOF.– We represent the action of the operator $\mathbf{L}_t^\varepsilon$ on the function $\varphi^\varepsilon(v,w,x)$ as follows:

$$\begin{aligned}\mathbf{L}_t^\varepsilon(v,w,x) &= \varepsilon^{-4}\mathbf{Q}\varphi(v,w) + \varepsilon^{-2}[\mathbf{Q}\varphi_2(v,w,x) + \mathbf{C}_0(x)\varphi(v,w)] \\ &+ \varepsilon^{-1}[\mathbf{Q}\varphi_3(v,w,x) + \frac{a}{\sqrt{t}}\mathbf{C}^0(x)\varphi(v,w)] \\ &+ \frac{1}{t}[\mathbf{Q}\varphi_4(v,w,x) + a\mathbf{C}_0(x)\varphi_2(v,w,x) \\ &+ \mathbf{C}^1(v,w,x)\varphi(v,w)] + \theta_L^\varepsilon(v,w,x)\varphi(v,w)]. \end{aligned} \quad [4.222]$$

Note that $\mathbf{Q}(v,w) = 0$ because the function $\varphi(v,w)$ is independent of x. Using the solvability condition

$$\mathbf{Q}\varphi_2(v,w,x) + \mathbf{C}_0(x)\varphi(v,w) = 0$$

and condition (BC1), we obtain the following representation:

$$\varphi_2(v,w,x) = R_0\mathbf{C}_0(x)\varphi(v,w). \quad [4.223]$$

Using the solvability condition

$$\mathbf{Q}\varphi_3(v,w,x) + \frac{a}{\sqrt{t}}\mathbf{C}^0(x)\varphi(v,w) = 0$$

for the singular perturbation problem [4.222] and condition (BC2), we get

$$\varphi_3(v,w,x) = \frac{a}{\sqrt{t}}R_0\mathbf{C}^0(x)\varphi(v,w). \quad [4.224]$$

The penultimate term in [4.222] gives the limit operator $\mathbf{L}(x)$:

$$\mathbf{Q}\varphi_4(v,w,x) + a\mathbf{C}_0(x)\varphi_2(v,w,x) + \mathbf{C}^1(v,w,x)\varphi(v,w) = \mathbf{L}(x)\varphi(v,w). \quad [4.225]$$

With regard to [4.224], relation [4.225] yields

$$\mathbf{Q}\varphi_4(v,w,x) + [a\mathbf{C}_0(x)R_0\mathbf{C}_0(x) + \mathbf{C}^1(v,w,x)]\varphi(v,w) = \mathbf{L}(x)\varphi(v,w). \quad [4.226]$$

The limit operator $\mathbf{L}$ is determined from the solvability condition for equation [4.226]:

$$\mathbf{L}\Pi = \Pi\mathbf{L}(x)\Pi. \qquad [4.227]$$

Using relations [4.218], [4.220] and [4.227], we obtain the limit operator $\mathbf{L}$ in the form [4.212].

COROLLARY 4.4.– *The presence of the factor $1/t$ in representation [4.221] means that, under the normalization $\tau = e^t$, the limit process $\eta(t) = \zeta(e^t)$ defined by the generator $\mathbf{L}$ is a diffusion process.*

PROOF OF THEOREM 4.5.– The weak convergence [4.211] is proved by using the model limit theorem.

Thus, the asymptotic normality of a continuous procedure of stochastic approximation in the Euclidean space R^d, $d > 1$, can be realized with additional technical complications.

4.4. Asymptotic normality of SAP in a semi-Markov environment

Now we consider the discrete procedure of stochastic approximation defined by the relation

$$u_{n+1}^{\varepsilon} = u_n^{\varepsilon} + \varepsilon^2 a_n^{\varepsilon} C(u_n^{\varepsilon}, x_n^{\varepsilon}), \ n \geq 0, \qquad [4.228]$$

where the regression function $C(u, x)$, $u \in R$, $x \in X$, satisfies the following condition for the existence of a global solution of the accompanying system:

$$\frac{du_x(t)}{dt} = C(u_x(t), x), \ x \in X.$$

A semi-Markov switching process $x(t), t \geq 0$, in the standard phase space $(X, \mathbf{X})$, is defined by the semi-Markov kernel

$$\mathbf{Q}(x, B, t) = P(x, B)G_x(t), \ x \in X, B \in X, t \geq 0, \qquad [4.229]$$

where the stochastic kernel $P(x, B)$ defines transition probabilities of the embedded Markov chain $x_n, n \geq 0$, namely

$$P(x, B) = Px_{n+1} \in B|x_n = x,$$

and the distribution functions $G_x(t)$, $x \in X, t \geq 0$, define Markov renewal moments $\tau_{n+1} = \tau_n + \theta_{n+1}$, $n \geq 0$, $\tau_0 = 0$:

$$G_x(t) = P\theta_{n+1} \leq t|x_n = x, \ x \in X, t \geq 0.$$

If the normalizing sequence a_n^ε, $n \geq 0$, satisfies the corresponding conditions and the semi-Markov process $x(t)$, $t \geq 0$, with stationary distribution $\pi(B)$, $B \in X$, is uniformly ergodic, then the discrete procedure of stochastic approximation defined by a solution of the evolution equation [4.228] converges with probability 1 to the equilibrium point u_0 of the averaged system

$$\frac{du(t)}{dt} = C(u(t)), \; C(u_0) = 0, \tag{4.230}$$

with the regression function

$$C(u) = \int_X \pi(dx)C(u, x).$$

Parallel with the discrete procedure of stochastic approximation [4.228], we consider the jump stochastic procedure

$$u^\varepsilon(t) = u_0 + \varepsilon^2 \sum_{n=0}^{\nu(t/\varepsilon^2)-1} a_n^\varepsilon C(u_n^\varepsilon, x_n^\varepsilon), \; t > 0, \tag{4.231}$$

where $\nu(t) := maxn : \tau_n \leq t$ is the counting process of the moments of jumps τ_n, $n \geq 1$.

The embedding of the discrete procedure of stochastic approximation [4.228] into the jump stochastic procedure [4.231] is determined by the relations

$$u_n^\varepsilon = u^\varepsilon(\tau_n^\varepsilon), \; a_n^\varepsilon = a(\tau_n^\varepsilon), \; \tau_n^\varepsilon := \varepsilon^2 \tau_n. \tag{4.232}$$

NOTE 4.4.– *The convergence of the procedure of stochastic approximation*

$$u^\varepsilon(t) \Rightarrow 0, \; t \to \infty \tag{4.233}$$

means that it is reasonable to study fluctuations with the following normalization:

$$v^\varepsilon(t) = \frac{\sqrt{t}u^\varepsilon(t)}{\varepsilon}. \tag{4.234}$$

For the discrete procedure of stochastic approximation [4.228], using [4.232] and [4.234], we obtain the normalized discrete procedure of stochastic approximation

$$\nu_n^\varepsilon = \frac{\sqrt{\tau_n^\varepsilon}u_n^\varepsilon}{\varepsilon}, \tag{4.235}$$

or, in the inverse form,

$$u_n^\varepsilon = \frac{\varepsilon v_n^\varepsilon}{\sqrt{\tau_n^\varepsilon}}. \tag{4.236}$$

Furthermore, without loss of generality, we assume that $u_0 = 0$, i.e. the following condition is satisfied:

(Y2) $C(0) = 0$.

We consider a standard procedure of stochastic approximation with the following normalizing function:

(Y3) $a(t) = a/t, t > 0$.

Additional conditions for the regression function are as follows:

(Y4) $C(u, \cdot) \in C^3(R)$.

In other words, the second derivative $C_u^n(u, \cdot)$ with respect to u satisfies the global Lipschitz condition

$$|C_u^n(u, x) - C_u^n(u', x)| \leq C|u - u'|,$$

where the constant C is independent of $x \in X$.

It is known (Korolyuk and Limnios 2005) that the uniform ergodicity of the semi-Markov process $x(t)$, $t \geq 0$, with semi-Markov kernel [4.229] determines the generator $\mathbf{Q}$ of the accompanying uniform ergodic Markov process $x^0(t)$, $t \geq 0$, namely,

$$\mathbf{Q}(x) = q(x) \int_X P(x, dy)[\varphi(y) - \varphi(x)],$$

where

$$q(x) := \frac{1}{g(x)}, \; g(x) := \int_0^\infty \bar{G}_x(t)dt, \; \bar{G}_x(t) := 1 - G_x(t).$$

Moreover, the generator $\mathbf{Q}$ is a reducibly invertible operator for which there exists a potential R_0 defined by the equation

$$R_0\mathbf{Q} = \mathbf{Q}R_0 = I - \Pi.$$

Here, the projector Π in the Banach space $\mathcal{B}(X)$ of real-valued functions with the supremum norm is defined by the relation

$$\Pi\varphi(x) := \hat{1}(x), \; \hat{\varphi} := \int_X \pi(dx)\varphi(x), \; \mathbf{1}(x) \equiv 1, \; x \in X.$$

Taking condition (Y4) into account, we use the Taylor formula for the regression function:

$$C(u,x) = C_0(x) + uC_1(x) + \frac{u^2}{2}C_2(u,x). \qquad [4.237]$$

Here, by definition,

$$C0(x) := C(0,x),\ C_1(x) := C'_u(u,x)|_{u=0},$$
$$C_2(u,x) := C''_u(\theta u,x),\ 0 \le \theta \le 1. \qquad [4.238]$$

Denote

$$c := -\int_X \pi(dx)q(x)C_1(x),\ b := ac - \frac{1}{2}.$$

Let the convergence conditions be satisfied for the procedure of stochastic approximation [4.228] in a semi-Markov environment. Namely, we assume that there exists a Lyapunov function $V(u)$, $u \in R$, that guarantees the exponential stability of system [4.230]:

(C1) $C(u)V'(u) \le -c_0V(u)$, $c_0 > 0$;

Also assume that the function $\tilde{C}(u,x) := C(u) - C(u,x)$, satisfies the following additional conditions:

(C2) $|\tilde{C}(u,x)V'(u)| \le c_1V(u)$, $c_1 > 0$;

$|C(u,x)[\tilde{C}(u,x)V'(u)]'| \le c_2V(u)$, $c_2 > 0$;

$|C(u,x)[C(u,x)[\tilde{C}(u,x)V'(u)]']'| \le c_3V(u)$, $c_3 > 0$.

Moreover, the distribution functions $G_x(t)$, $x \in X$, $t \ge 0$, satisfy the Cramer condition uniformly in $x \in X$:

(C3) $\sup\limits_{x\in X} \int\limits_0^\infty e^{ht}\bar{G}_x(dt) \le H < +\infty$, $h > 0$.

We now formulate the theorem on asymptotic normality.

We consider the following additional conditions:

(D1) $\rho := -2\int\limits_X \pi(dx)[\tilde{C}_0(x)R_0\tilde{C}_0(x) + \frac{1}{2}q(x)C_0^2(x)] > 0$,

where $\tilde{C}_0(x) := q(x)C_0(x)$,

(D2) $b := ac - \frac{1}{2} > 0$.

THEOREM 4.6.– *Under conditions C1–C3 for the convergence [4.233] of the procedure of stochastic approximation [4.228] in a semi-Markov medium and additional conditions (D1) and (D2), the weak convergence*

$$v^\varepsilon(t) \Rightarrow \zeta(t), \ \varepsilon \to 0, \qquad [4.239]$$

takes place on every finite interval $0 < t_0 \leq t \leq T$. The limit process $\zeta(t)$, $t \geq 0$, is the Ornstein–Uhlenbeck diffusion process defined by the operator

$$\mathbf{L}\varphi(v) = -bv\varphi'(v) + \frac{a^2}{2}\rho\varphi''(v). \qquad [4.240]$$

NOTE 4.5.– *Condition (D1) guarantees the diffusivity of the limit process $\zeta(t)$, $t \geq 0$, and condition $D2$ means the ergodicity of the process $\zeta(t)$, $t \geq 0$, with stationary normal distribution $N(0, \sigma_0^2)$, $\sigma_0^2 = \rho/2b$.*

Taking into account the weak convergence [4.239], we arrive at the following conclusion:

Conclusion 1. Under the conditions of the theorem, the normalized procedure of stochastic approximation $v^\varepsilon(t)$ has the asymptotically normal distribution $N(0, \sigma_0^2)$, i.e.

$$v^\varepsilon(t) \Rightarrow \nu, \ \varepsilon \to 0, \ t \to \infty,$$

where the random variable ν belongs to $N(0, \sigma_0^2)$.

Furthermore, we investigate the properties of the normalized discrete procedure of stochastic approximation [4.235].

Let us consider the Taylor formula

$$\sqrt{t + \varepsilon^2 s} = \sqrt{t}(1 + \frac{\varepsilon^2 s}{2t} + \varepsilon^2 h^\varepsilon(s)), \qquad [4.241]$$

where $h^\varepsilon(s) > 0$, $\varepsilon \to 0$, uniformly on every finite segment $s \in [0, S_0]$, $S_0 < \infty$.

Thus, the increment of the function $\sqrt{t}$ for the increment $\varepsilon^2 s$ of the argument has the form

$$\sqrt{t + \varepsilon^2 s} - \sqrt{t} = \frac{\varepsilon^2 s}{2t} + \varepsilon^2 h^\varepsilon(s))\sqrt{t}. \qquad [4.242]$$

LEMMA 4.19.– *The normalized discrete procedure of stochastic approximation [4.235] satisfies the relation*

$$v_{n+1}^\varepsilon = v_n^\varepsilon + \varepsilon\frac{a}{\sqrt{\tau_n^\varepsilon}}C(\frac{\varepsilon v_n^\varepsilon}{\sqrt{\tau_n^\varepsilon}}, x_n^\varepsilon) + \varepsilon^2\frac{\theta_n^\varepsilon}{2\tau_n^\varepsilon}v_n^\varepsilon + \varepsilon^2 H^\varepsilon(\theta_{n+1}), \qquad [4.243]$$

where the remainder $H^\varepsilon(\theta_{n+1})$ is such that $\mathbb{E}H^\varepsilon(\theta_{n+1}) \to 0$, $\varepsilon \to 0$.

PROOF.– For the increment $\Delta v_n^\varepsilon := v_{n+1}^\varepsilon - v_n^\varepsilon$ of the normalized discrete procedure of stochastic approximation [4.235], we have the representation

$$\Delta v_n^\varepsilon = \frac{\sqrt{\tau_{n+1}^\varepsilon}u_{n+1}^\varepsilon - \sqrt{\tau_n^\varepsilon}u_n^\varepsilon}{\varepsilon},$$

or

$$\Delta v_n^\varepsilon = \frac{\sqrt{\tau_{n+1}^\varepsilon}\Delta u_n^\varepsilon + (\sqrt{\tau_{n+1}^\varepsilon} - -\sqrt{\tau_n^\varepsilon})u_n^\varepsilon}{\varepsilon}. \qquad [4.244]$$

According to [4.228] and [4.232], the increment Δu_n^ε is determined by the relation

$$\Delta u_n^\varepsilon = \varepsilon^2 a_n^2 C(u_n^\varepsilon, x_n^\varepsilon).$$

Taking [4.236] and [4.232] into account, we obtain the representation

$$\Delta u_n^\varepsilon = \varepsilon^2 \frac{a}{\tau_n^\varepsilon} C(\frac{\varepsilon v_n^\varepsilon}{\sqrt{\tau_n^\varepsilon}}, x_n^\varepsilon). \qquad [4.245]$$

Substituting [4.245] into [4.244] and taking [4.241] and [4.242] into account, we get

$$\Delta v_n^\varepsilon = \frac{\sqrt{\tau_{n+1}^\varepsilon}\varepsilon^2 a C(\varepsilon v_n^\varepsilon/\sqrt{\tau_{n+1}^\varepsilon}, x_n^\varepsilon)/\tau_{n+1}^\varepsilon}{\varepsilon}$$

$$+ \frac{\varepsilon^2(\theta_{n+1}/(2\sqrt{\sqrt{\tau_n^\varepsilon}} + \sqrt{\tau_n^\varepsilon}h^\varepsilon(\theta_{n+1})\varepsilon v_n^\varepsilon/(2\sqrt{\tau_n^\varepsilon})\varepsilon}{}$$

$$= \varepsilon \frac{a\sqrt{\tau_{n+1}^\varepsilon}}{\tau_n^\varepsilon} C(\frac{\varepsilon v_n^\varepsilon}{\sqrt{\tau_n^\varepsilon}}, x_n^\varepsilon) + \varepsilon^2 \frac{\theta_{n+1}}{2\tau_n^\varepsilon} v_n^\varepsilon + \varepsilon^2 h_1^\varepsilon(\theta_{n+1})$$

$$= \varepsilon \frac{a}{\sqrt{\tau_n^\varepsilon}} C(\frac{\varepsilon v_n^\varepsilon}{\sqrt{\tau_n^\varepsilon}}, x_n^\varepsilon) + \varepsilon^2 \frac{\theta_{n+1}}{2\tau_n^\varepsilon} v_n^\varepsilon + \varepsilon^2 H^\varepsilon(\theta_{n+1}),$$

where $h_1^\varepsilon(\theta_{n+1})$ possesses the same properties as $h^\varepsilon(\theta_{n+1})$ in [4.241].

Using the last results, we obtain the statement of Lemma 4.16.

Conclusion 2. The leading term of the increment of the normalized discrete procedure of stochastic approximation [4.235] has the form

$$\Delta v_n^\varepsilon = \varepsilon \frac{a}{\sqrt{\tau_n^\varepsilon}} C(\frac{\varepsilon v_n^\varepsilon}{\sqrt{\tau_n^\varepsilon}}, x_n^\varepsilon) + \varepsilon^2 \frac{\theta_{n+1}}{2\tau_n^\varepsilon} v_n^\varepsilon. \qquad [4.246]$$

Using [4.237], we obtain the following corollary:

COROLLARY 4.5.– *In notation [4.238], increment [4.246] has the form*

$$\Delta v_n^\varepsilon = \varepsilon \frac{a}{\sqrt{\tau_n^\varepsilon}} C_0(x) + \varepsilon^2 [\frac{a v_n^\varepsilon}{\tau_n^\varepsilon} C_1(x) + \frac{\theta_{n+1}}{2\tau_n^\varepsilon} v_n^\varepsilon] + \varepsilon^2 h^\varepsilon(v, x), \qquad [4.247]$$

where $h^\varepsilon(v,x)$ is a negligible term such that $|h^\varepsilon(v,x)| \to 0$, $\varepsilon \to 0$, $t \geq t_0 > 0$.

Now we can use the compensating operator of the extended Markov renewal process

$$v_n^\varepsilon = v^\varepsilon(\tau_n^\varepsilon),\ x_n^\varepsilon := x(\frac{\tau_n^\varepsilon}{\varepsilon^2}) = x(\tau_n),\ \tau_n^\varepsilon = \varepsilon^2\tau_n,\ n \geq 0, \qquad [4.248]$$

defined by the relation

$$\mathbf{L}_t^\varepsilon\varphi(v,x,t) = \varepsilon^{-2}q(x)[\mathbb{E}[\varphi(v_{n+1}^\varepsilon, x_{n+1}^\varepsilon, \tau_{n+1}^\varepsilon)|v_n^\varepsilon = v,\ x_n^\varepsilon = x,\ \tau_n^\varepsilon = t] \\ - \varphi(v,x,t)]. \qquad [4.249]$$

Consider the semigroup

$$\Gamma_s\varphi(v) = \varphi(v + \frac{v}{2t}s),\ s \geq 0, \qquad [4.250]$$

with the generating operator $\Gamma\varphi(v) = \frac{v}{2t}\varphi'(v)$ and the jump operator

$$\mathbf{D}^\varepsilon\varphi(v) = \varphi(v + \varepsilon\frac{a}{\sqrt{t}}C\frac{\varepsilon v}{\sqrt{t}}),\ t > 0. \qquad [4.251]$$

LEMMA 4.20.– *The compensating operator [4.249] has the analytical representation*

$$\mathbf{L}_t^\varepsilon\varphi(v,x) = \varepsilon^{-2}\mathbf{Q}\varphi(v,x) + \varepsilon^{-2}q(x)\int_0^\infty G_x(ds)[\Gamma_{\varepsilon^2 s}\mathbf{D}^\varepsilon - I]\mathbf{P}\varphi(v,x), \qquad [4.252]$$

where

$$\mathbf{P}\varphi(x) := \int_X P(x,dy)\varphi(y).$$

PROOF.– The proof is based on the determination of the conditional mathematical expectation in [4.249] with regard to the relation

$$v_{n+1}^{\varepsilon+1} = v_n^{\varepsilon+1} + v^\varepsilon(\varepsilon^2\theta_{n+1}).$$

Representation [4.249] has the following operator form:

$$\mathbf{L}_t^\varepsilon\varphi(v,x) = \varepsilon^{-2}\mathbf{Q}\varphi(v,x) + \varepsilon^{-2}q(x)[\int_0^\infty G_x(ds)\mathbf{P}\varphi(v + \Delta v^\varepsilon, x) - \varphi(v,x)], \qquad [4.253]$$

where the increment Δv^ε consists of the jump component defined by operator [4.251] and the continuous component defined by semigroup [4.250]. Their sum gives the leading increment [4.246] of the normalized discrete procedure of stochastic approximation.

To study the asymptotics of the compensating operator, we consider the asymptotics of the increment of the evolution of $\varphi(v,x)$. Denoting $s = \theta^\varepsilon_{n+1}$, $x = x^\varepsilon_n$, $t = \tau^\varepsilon_n$ and $v = v^\varepsilon_n$ and taking [4.247] into account, we get

$$\begin{aligned}\Delta\varphi(v,x) :&= \varphi(v+\Delta v^\varepsilon, x) - \varphi(v,x)\\ &= \varphi(v + \varepsilon\frac{a}{\sqrt{t}}C_0(x) + \varepsilon^2\frac{v}{t}[aC_1(x) + \frac{s}{2}]\\ &\quad + \varepsilon^2 h^\varepsilon(v,s), x) - \varphi(v,x).\end{aligned}$$

Thus, for the corresponding smoothness of $\varphi(v,\cdot) \in C^3(R)$, we finally obtain

$$\Delta\varphi(v,x) = [\varepsilon\frac{a}{\sqrt{t}}C_0(x) + \varepsilon^2\frac{v}{t}b(x,s)]\varphi'(v,x) + \varepsilon^2\frac{a^2}{2t}C_0^2\varphi''(v,x) + \varepsilon^2 h^\varepsilon(v,x),$$

where

$$b(x,s) = aC_1(x) + \frac{s}{2}$$

and $h^\varepsilon(v,x)$ is such that $|h^\varepsilon(v,x))| \to 0$, $\varepsilon \to 0$, $t \geq t_0 > 0$.

Taking the last result into account, we obtain [4.252] from [4.253].

LEMMA 4.21.– *On the test functions $\varphi(v,x) \in C^3(R)$, $x \in X$, the compensating operator [4.252] has the asymptotic representation*

$$\begin{aligned}\mathbf{L}^\varepsilon_t\varphi(v,x) &= \varepsilon^{-2}\mathbf{Q}\varphi(v,x) + \varepsilon^{-1}\frac{1}{\sqrt{t}}\mathbf{Q}_1(x)\mathbf{P}\varphi(v,x)\\ &\quad + \frac{1}{t}\mathbf{Q}_2(x)\mathbf{P}\varphi(v,x) + \theta^\varepsilon(x)\varphi(v,x),\end{aligned} \qquad [4.254]$$

where

$$\mathbf{Q}_1(x)\varphi(v) = aq(x)C_0(x)\varphi'(v),$$

$$\mathbf{Q}_2(x)\varphi(v) = vb(x)q(x)\varphi'(x) + \frac{a^2}{2}q(x)C_0^2(x)\varphi''(v),$$

$$b(x) = aC_1(x) + \frac{1}{2},$$

and the remainder $\theta^\varepsilon(x)\varphi(v,x)$ is such that

$$\|\theta^\varepsilon(x)\varphi(v,x)\| \to 0,\ \varepsilon \to 0,\ \varphi(v,\cdot) \in C^3(R)$$

PROOF.– Consider the second term in [4.252], denoting

$$\mathbf{L}_0^\varepsilon \varphi(v,x) = \varepsilon^{-2} q(x) \int_0^\infty G_x(ds)[\Gamma_{\varepsilon^2 s}\mathbf{D}^\varepsilon - I]\mathbf{P}\varphi(v,x)$$
$$= [\mathbf{L}_\Gamma^\varepsilon + \mathbf{L}_D^\varepsilon + \mathbf{L}_{\Gamma D}^\varepsilon]\varphi(v,x),$$

where

$$\mathbf{L}_\Gamma^\varepsilon \varphi(v,x) := \varepsilon^{-2} q(x) \int_0^\infty G_x(ds)[\Gamma_{\varepsilon^2 s} - I]\mathbf{P}\varphi(v,x),$$

$$\mathbf{L}_D^\varepsilon \varphi(v,x) := \varepsilon^{-2} q(x) \int_0^\infty G_x(ds)[\mathbf{D}^\varepsilon - I]\mathbf{P}\varphi(v,x).$$

$$\mathbf{L}_{\Gamma\mathbf{D}}^\varepsilon \varphi(v,x) := \varepsilon^{-2} q(x) \int_0^\infty G_x(ds)[\Gamma_{\varepsilon^2 s} - I][\mathbf{D}^\varepsilon - I]\mathbf{P}\varphi(v,x).$$

Consider the properties of each introduced operator. Taking into account the equation for the semigroup

$$\Gamma_{\varepsilon^2 s} - I = \Gamma \int_0^{\varepsilon^2 s} \Gamma_t dt$$

and integrating by parts, we get

$$\mathbf{L}_\Gamma^\varepsilon \varphi(v,x) = [\Gamma\mathbf{P} + \theta_\Gamma^\varepsilon(x)]\varphi(v,x). \quad [4.255]$$

Using the expansion according to the Taylor formula in [4.251], we get

$$\mathbf{L}_D^\varepsilon \varphi(v,x) = [\varepsilon^{-1} D_1(x)\mathbf{P} + D_2(x)\mathbf{P}]\varphi(v,x) + \theta_D^\varepsilon(x)\varphi(v,x), \quad [4.256]$$

where

$$D_1(x)\varphi(v) = \frac{a}{\sqrt{t}} C_0(x)\varphi'(v),$$

$$D_2(x)\varphi(v) = \frac{a}{t}[C_1(x)\varphi'(v) + \frac{a}{2}C_0^2(x)\varphi''(v)].$$

It is easy to verify that

$$\mathbf{L}_{\Gamma D}^\varepsilon \varphi(v,x) = \theta_{\Gamma D}^\varepsilon(x)\varphi(v,x). \quad [4.257]$$

Here, the operators $\theta_*^\varepsilon(x)$ are negligible on the class of the test functions $\varphi(v,\cdot) \in C^k(R)$, $k \geq 3$.

Summing up the results of the asymptotic decompositions [4.255]–[4.257], we obtain the statement of Lemma 4.18.

PROOF OF THEOREM 4.6.– The proof of the theorem on the asymptotic normality of the normalized procedure of stochastic approximation [4.235] is based on the application of the solution of the singular perturbation problem to the truncated compensating operator

$$\mathbf{L}_0^\varepsilon \varphi(v,x) = \varepsilon^{-2}\mathbf{Q}\varphi(v,x) + \varepsilon^{-1}\frac{1}{\sqrt{t}}\mathbf{Q}_1\mathbf{P}\varphi(v,x) + \frac{1}{t}\mathbf{Q}_2\mathbf{P}\varphi(v,x)$$

with representation [4.254].

Using the perturbed test function

$$\varphi^\varepsilon(v,x,t) = \varphi(v) + \varepsilon\frac{1}{\sqrt{t}}\varphi_1(v,x) + \varepsilon^2\frac{1}{t}\varphi_2(v,x)$$

and relation [4.24] from Lemma 4.12, we get

$$\begin{aligned}\mathbf{L}_0^\varepsilon \varphi^\varepsilon(v,x,t) &= [\varepsilon^{-2}\mathbf{Q} + \varepsilon^{-1}\frac{1}{\sqrt{t}}\mathbf{Q}_1(x)\mathbf{P} + \frac{1}{t}\mathbf{Q}_2(x)\mathbf{P}] \\ &= [\varphi(v) + \varepsilon\frac{1}{\sqrt{t}}\varphi_1(v,x) + \varepsilon^2\frac{1}{t}\varphi_2(v,x)] \\ &= \frac{1}{t}[\mathbf{Q}\varphi_2(v,x) + (\mathbf{Q}_1(x)\mathbf{P}R_0\mathbf{Q}_1(x) \\ &\quad + \mathbf{Q}_2(x))\varphi(v)] + \theta_L^\varepsilon(x,t)\varphi(x) \\ &= \frac{1}{t}\mathbf{L}\varphi(v) + \theta_L^\varepsilon(x,t)\varphi(x),\end{aligned}$$

where the operator $\mathbf{L}$ is such that

$$\mathbf{L}\Pi = \Pi\mathbf{Q}_1(x)\mathbf{P}R_0\mathbf{Q}_1(x)\Pi + \Pi\mathbf{Q}_2(x)\Pi. \qquad [4.258]$$

Carrying out calculations according to relation [4.258], we obtain the generator [4.240] of the limit diffusion process. To complete the proof of the theorem, we use the scheme of the proof of Theorem 6.6 (Chabanyuk 2006).

Conclusion 3. An analogous result on asymptotic normality for a jump procedure of stochastic approximation in the Euclidean space R^d, $d > 1$, can be obtained with additional technical complications.

5

Stochastic Optimization Procedure

In this chapter, we consider a stochastic optimization procedure under the average scheme in the case of direct dependence on the regression function of a switching Markov process that satisfies the Lipschitz condition. We also prove the asymptotical normality, the stochastic optimization procedure which is described by the Ornstein–Uhlenbeck process.

5.1. SOP in the average scheme

5.1.1. *Convergence SOP*

We investigate a continuous stochastic optimization procedure for finding the point of maximum $u_0 \in R^d$ of the regression function $C(u,x)$, which directly depends on the influence of the external environment on Markov switchings.

Let $C(u,x), u \in R^d$, be a regression function that has its unique maximum at a point $u_0, u_0 \in R^d$. The second component x of the regression function characterizes the impact of external factors, which are described by a uniformly ergodic Markov process $x(t), t \geq 0$, on the measured phase space of states $(X, \mathbf{X})$. The generator of the Markov process can be determined by the relation

$$Q\varphi(x) = q(x) \int_X P(x,dy) \left[\varphi(y) - \varphi(x)\right],$$

on the Banach space $\mathbf{B}(X)$ of real-valued bounded continuous functions $\varphi(x), x \in X$, with the norm

$$\|\varphi(x)\| = \sup_{x \in X} |\varphi(x)|,$$

where $P(x,B), x \in X, B \in \mathbf{X}$ is the stochastic kernel (Anisimov (2008)), $q(x) = g^{-1}(x)$, $g(x) = E\theta_x$, θ_x is the staying time of the Markov process at state x.

The stationary distribution $\pi(B), B \in \mathbf{X}$ of the Markov process $x(t), t \geq 0$, is determined by the relation

$$\pi(dx)q(x) = q\rho(dx),\ q = \int_X \pi(dx)q(x),$$

where $\rho(dx)$ is the stationary distribution of the embedded Markov chain. We also determine the potential

$$R_0 = \Pi - (\Pi + Q)^{-1},$$

for the generator Q of the Markov process $x(t), t \geq 0$, where $\Pi\varphi(x) = \int_X \pi(dx)\varphi(x)$ is the projector onto the zero-subspace of the operator Q: $N_Q = \{\varphi : Q\varphi = 0\}$ (Anisimov 2008).

A continuous stochastic optimization procedure for the regression function $C(u, x)$ in an ergodic Markov environment is given by the stochastic differential equation:

$$du^\varepsilon(t) = a(t)\nabla_{b(t)}C(u^\varepsilon(t), x(t/\varepsilon))dt, \qquad [5.1]$$

where

$$\nabla_{b(t)}C(u, x) = \Big\{\frac{C(u_i^+, x) - C(u_i^-, x)}{2b(t)}, i \in \overline{1, d}\Big\},$$

$$u_i^\pm = u_i \pm b(t)e_i, i = \overline{1, d}, e_i = (0, \ldots, 1, 0, \ldots, 0)$$

Furthermore, we consider the evolutionary system

$$\frac{du}{dt} = C'(u), \tilde{N}' = \operatorname{grad} C(u) = \Big\{\frac{\partial C(u)}{\partial u_i}, i \in \overline{1, d}\Big\}. \qquad [5.2]$$

for the averaged regression function

$$C(u) = \int_X \pi(dx)C(u, x).$$

THEOREM 5.1.– *Let the Lyapunov function $V(u), u \in R^d$, for the averaged system [5.2] satisfy the following conditions:*

C1: (exponential stability of the evolutionary system [5.2])

$$C'(u)V'(u) \leq -c_0 V(u), c_0 > 0;$$

C2: $|V'(u)| \leq c_1(1 + V(u)), c_1 > 0;$

C3: $|\nabla_{b(t)}C(u,x)R_0[\nabla_{b(t)}\tilde{C}(u,x)V'(u)]'| \le c_2(1+V(u)), c_2 > 0,$

$$\tilde{C}(u,x) := C(u,x) - C(u).$$

The regression function $C(u)$ at u satisfies the global Lipschitz condition:

C4: $|\nabla_{b(t)}C(u) - C'(u)| \le c_3 b(t), c_3 > 0.$

The functions $a(t), b(t), t \ge 0$ satisfy the convergence conditions of the stochastic optimization procedure:

C5: $$\int_0^{+\infty} a(t)dt = \infty, \int_0^{+\infty} a(t)b(t)dt < \infty, a(t) > 0, b(t) > 0.$$

Then, for every initial value $u^\varepsilon(0) \in R^d$, the stochastic optimization procedure [5.1] for all $\varepsilon \le \varepsilon_0$, where ε_0 is small enough, converges with probability 1 to the point of maximum u_0.

THEOREM 5.2.– *We assume the existence of the Lyapunov function $V(u), u \in R^d$ for the average system [5.2], satisfying the following conditions:*

C1: (exponential stability of the evolutionary system [5.2])

$$C'(u)V'(u) \le -c_0 V(u), c_0 > 0;$$

C2: $|V'(u)| \le c_1(1+V(u)), c_1 > 0;$

C3: $|\nabla_{b(t)}C(u,x)R_0[\nabla_{b(t)}\tilde{C}(u,x)V'(u)]'| \le c_2(1+V(u)), c_2 > 0,$

$\tilde{C}(u,x) := C(u,x) - C(u).$

The regression function $C(u)$ at evolution u satisfies the global Lipschitz condition

C4: $|\nabla_{b(t)}C(u) - C'(u)| \le c_3 b^2(t), c_3 > 0.$

The functions $a(t), b(t), t \ge 0$ satisfy the convergence conditions of the stochastic optimization procedure

C5: $$\int_0^{+\infty} a(t)dt = \infty, \int_0^{+\infty} a(t)b^2(t)dt < \infty, a(t) > 0, b(t) > 0.$$

Then, for each initial value $u^\varepsilon(0) \in R^d$, the stochastic optimization procedure [5.1] for all $\varepsilon \le \varepsilon_0$, where ε_0 is small enough, converges with probability 1 to the point of maximum u_0.

To prove Theorem 5.1, we first construct the generator of the stochastic optimization procedure and then investigate its asymptotic representation

LEMMA 5.1.– *The generator of the Markov process*

$$u^\varepsilon(t), x_t^\varepsilon = x(t/\varepsilon), t \geq 0 \tag{5.3}$$

on the Banach space $\boldsymbol{B}(R^d, X)$ *of real-valued functions* $\varphi(u,x) \in C^{2,0}(R^d, X)$ *has the form*

$$\boldsymbol{L}_t^\varepsilon \varphi(u,x) = \varepsilon^{-1} Q\varphi(u,x) + a(t)\nabla_{b(t)}\boldsymbol{C}(u)\varphi(u,x),$$

where

$$\nabla_{b(t)}\boldsymbol{C}(u)\varphi(u,x) = \nabla_{b(t)}C(u,x)\varphi'_u(u,x).$$

PROOF.– We consider the conditional expectation of the test functions $\varphi(u,x) \in C^{2,0}(R^d, X)$

$$E[\varphi(u^\varepsilon(t+\Delta), x_{t+\Delta}^\varepsilon)\,|u^\varepsilon(t) = u, x_t^\varepsilon = x]$$

$$E_{u,x}\varphi(u^\varepsilon(t+\Delta), x_{t+\Delta}^\varepsilon)$$

$$= E_{u,x}\varphi(u + \int_t^{t+\Delta} a(s)\nabla_{b(s)}C(u^\varepsilon(s), x(s/\varepsilon))ds, x) I(\theta > \varepsilon^{-1}\Delta)$$

$$+ E_{u,x}\varphi(u + \int_t^{t+\Delta} a(s)\nabla_{b(s)}C(u^\varepsilon(s), x(s/\varepsilon))ds, x_{t+\Delta}^\varepsilon) I(\theta \leq \varepsilon^{-1}\Delta) + o(\Delta).$$

The distribution function of the staying time θ_x at state x has an exponential distribution, i.e. we have decompositions

$$I(\theta_x > \varepsilon^{-1}\Delta) = e^{-\varepsilon^{-1}q(x)\Delta} = 1 - \varepsilon^{-1}q(x)\Delta + o(\Delta)$$

and

$$I(\theta_x \leq \varepsilon^{-1}\Delta) = 1 - e^{-\varepsilon^{-1}q(x)\Delta} = \varepsilon^{-1}q(x)\Delta + o(\Delta),$$

where $q(x)$ is the intensivity.

Then, by the Taylor formula on the test function $\varphi(u,x) \in C^{2,0}(R^d, X)$, we obtain

$$\varphi(u + \int_t^{t+\Delta} a(s)\nabla_{b(s)}C(u^\varepsilon(s), x(s/\varepsilon))ds, x)$$

$$= \varphi(u,x) + \varphi'_u(u,x)\int_t^{t+\Delta} a(s)\nabla_{b(s)}C(u^\varepsilon(s), x(s/\varepsilon))ds$$

$$= \varphi(u,x) + \nabla_{b(t)}C(u,x)\varphi'_u(u,x)a(t)\Delta + o(\Delta).$$

Therefore,

$$\begin{aligned}
&E_{u,x}[\varphi(u,x) + a(t)\nabla_{b(t)}C(u,x)\varphi'_u(u,x)\Delta][1 - \varepsilon^{-1}q(x)\Delta + o(\Delta)] \\
&\quad + E_{u,x}[\varphi(u,x^\varepsilon_{t+\Delta}) + a(t)\nabla_{b(t)}C(u,x)\varphi'_u(u,x^\varepsilon_{t+\Delta})\Delta][\varepsilon^{-1}q(x)\Delta + o(\Delta)] \\
&= \varphi(u,x) + a(t)\nabla_{b(t)}C(u,x)\varphi'_u(u,x)\Delta - \varepsilon^{-1}q(x)E_{u,x}\varphi(u,x)\Delta \\
&\quad - \varepsilon^{-1}q(x)(a(t)E_{u,x}\nabla_{b(t)}C(u,x)\varphi'_u(u,x)\Delta^2) + o(\Delta) \\
&\quad + \varepsilon^{-1}q(x)E_{u,x}\varphi(u,x^\varepsilon_{t+\Delta})\Delta + o(\Delta) \\
&= \varphi(u,x) + a(t)\nabla_{b(t)}C(u,x)\varphi'_u(u,x)\Delta + \\
&\quad + \varepsilon^{-1}q(x)\Delta\int_X P(x,dy)[\varphi(u,y) - \varphi(u,x)] + o(\Delta).
\end{aligned} \qquad [5.4]$$

From [5.4], we obtain the generator of the Markov process [5.3]

$$\begin{aligned}
\boldsymbol{L}^\varepsilon_t\varphi(u,x) &= \lim_{\Delta\to 0}\frac{1}{\Delta}E_{u,x}(\varphi(u^\varepsilon(t+\Delta),x^\varepsilon_{t+\Delta}) - \varphi(u,x)) \\
&= \varepsilon^{-1}Q\varphi(u,x) + a(t)\nabla_{b(t)}C(u,x)\varphi'_u(u,x).
\end{aligned}$$

Furthermore, let the perturbed Lyapunov function have the form

$$V^\varepsilon(u,x) = V(u) + \varepsilon a(t)V_1(u,x), \qquad [5.5]$$

where $V(u) \in C^3(R^d)$ is the Lyapunov function for the average system [5.2].

LEMMA 5.2.– *The generator $\boldsymbol{L}^\varepsilon_t$ on the perturbed Lyapunov function $V^\varepsilon(u,x)$ has the asymptotic representation:*

$$\boldsymbol{L}^\varepsilon_t V^\varepsilon(u,x) = a(t)\boldsymbol{L}V(u) + \varepsilon a^2(t)\Theta_t(x)V(u),$$

where the limit generator has the form

$$\boldsymbol{L}V(u) = \nabla_{b(t)}C(u)V'(u),$$

and the remainder term

$$\Theta_t(x)V(u) = \nabla_{b(t)}C(u,x)R_0[\nabla_{b(t)}\tilde{C}(u,x)V'(u)]'.$$

PROOF.– The generator $\boldsymbol{L}^\varepsilon_t$ on the perturbed Lyapunov function [1.2] has the representation

$$\begin{aligned}
\boldsymbol{L}^\varepsilon_t V^\varepsilon(u,x) &= \left[\varepsilon^{-1}Q + a(t)\nabla_{b(t)}\boldsymbol{C}(x)\right]\left[V(u) + \varepsilon a(t)V_1(u,x)\right] \\
&= \varepsilon^{-1}QV(u) + a(t)QV_1(u,x) + a(t)\nabla_{b(t)}\boldsymbol{C}(x)V(u) \\
&\quad + \varepsilon a^2(t)\nabla_{b(t)}\boldsymbol{C}(x)V_1(u,x).
\end{aligned} \qquad [5.6]$$

Taking into account the solvability conditions of $QV(u) = 0$ for the singular perturbation problem [5.6], we obtain

$$a(t)QV_1(u,x) + a(t)\nabla_{b(t)}\boldsymbol{C}(x)V(u)$$
$$= a(t)\Big[QV_1(u,x) + \nabla_{b(t)}\boldsymbol{C}(x)V(u)\Big]$$
$$= a(t)\boldsymbol{L}V(u),$$

where the limit generator $\boldsymbol{L}$ has the representation

$$\boldsymbol{L}V(u) = \nabla_{b(t)}C(u)V'(u).$$

So, for perturbation $V_1(u,x)$, we have

$$V_1 = R_0\nabla_{b(t)}\tilde{C}(u,x)V(u).$$

Furthermore, for the remainder term from [5.6], we obtain

$$\varepsilon a^2(t)\nabla_{b(t)}\mathbf{C}(x)V_1(u,x)$$
$$= \varepsilon a^2(t)\nabla_{b(t)}\mathbf{C}(x)\big[R_0\nabla_{b(t)}\tilde{C}(u,x)V(u)\big]$$
$$= \varepsilon a^2(t)\nabla_{b(t)}C(u,x)R_0\big[\nabla_{b(t)}\tilde{C}(u,x)V'(u)\big]'.$$

Therefore, the remainder term has the form

$$\theta_t(x)V(u) = \nabla_{b(t)}C(u,x)R_0\big[\nabla_{b(t)}\tilde{C}(u,x)V'(u)\big]'.$$

Thus,

$$\boldsymbol{L}_t^\varepsilon V^\varepsilon(u,x) = a(t)\boldsymbol{L}V(u) + \varepsilon a^2(t)\Theta_t(x)V(u).$$

We now prove the theorem.

By condition $C1$ of the theorem, we have the estimation

$$\nabla_{b(t)}C(u)V'(u) \le -c_0V(u) - [C'(u) - \nabla_{b(t)}C(u)]V'(u),$$

and under C2–C5, we obtain the boundedness of the remainder term $\theta_t(x)V(u)$ in the form

$$|\theta_t(x)V(u)| \le c(1 + V(u)).$$

Then

$$\boldsymbol{L}_t^\varepsilon V^\varepsilon(u,x) \le -a(t)c_0V(u) + \varepsilon(a^2(t)c_4 - c_5a(t)b(t))(1 + V(u)). \qquad [5.7]$$

By [5.7] and the theorem by Nevelson and Khasminsky about convergence of the stochastic approximation procedure, we obtain the statement of Theorem 5.1.

PROOF OF THEOREM 5.2.– We provide it by the scheme of proving [5.1] by taking into account conditions $C4$ and $C5$ of Theorem 5.2, i.e. the estimations

$$\boldsymbol{L}_t^\varepsilon V^\varepsilon(u,x) \le -a(t)c_0V(u) + \varepsilon(a^2(t)c_4 - c_5a(t)b^2(t))(1 + V(u)).$$

5.1.2. *Asymptotical normality of Stochastic optimization procedure*

Furthermore, we investigate the asymptotic normality of the scalar stochastic optimization procedure with Markov switchings of the regression function under the classical scheme of normalizing at time with a small parameter.

Let $C(u,x)$, $u \in R$, be the regression function that has its unique maximum at a point u_0, $u_0 \in R$.

The component x characterizes the impact of external factors, described by a uniformly ergodic Markov process $x(t)$, $t \geq 0$, on the measured phase space of states $(X, \boldsymbol{X})$ Korolyuk and Limnios (2005). The generator of the MP is given by the relation:

$$Q\varphi(x) = q(x) \int_X P(x, dy) \left[\varphi(y) - \varphi(x)\right],$$

where $P(x, B), x \in X, B \in X$ is a stochastic kernel.

Furthermore, we consider a continuous stochastic optimization procedure for finding the point u_0 under the average scheme, which is given by the evolutionary differential equation:

$$\frac{du^\varepsilon(t)}{dt} = a(t)\nabla_{b(t)} C(u^\varepsilon(t), x(t/\varepsilon^2)) \qquad [5.8]$$

where

$$\nabla_{b(t)} C(u,x) = \frac{C(u+b(t),x) - C(u-b(t),x)}{2b(t)},$$

with normalizing functions

$$b(t) = \frac{b}{\sqrt{t}},\ b > 0$$

and

$$a(t) = \frac{a}{t},\ a > 0.$$

Asymptotical normality SOP is investigated under the exponential stability conditions for the averaged system

$$\frac{du}{dt} = C'(u), \qquad [5.9]$$

where

$$C(u) = \int_X \pi(dx) C(u,x)$$

is the averaged regression function with a stationary distribution $\pi(B),\ B \in X$.

We assume that the convergence conditions of the SOP in the average scheme to the equilibrium point u_0 of the averaged system [5.9] are fulfilled.

The normalized SOP has the form

$$v^{\varepsilon}(t) = \frac{\sqrt{t}u^{\varepsilon}(t)}{\varepsilon}. \qquad [5.10]$$

THEOREM 5.3.– *We consider the regression function $C(u, \cdot) \in C^3(R)$, and the following conditions hold true:*

$$C1 : \rho^2 := 2 \int_X \pi(dx)C'(0,x)R_0C'(0,x) > 0;$$
$$C2 : C_2 := \int_X \pi(dx)C''(0,x) < 0;$$
$$C3 : b := aC_2 + \tfrac{1}{2} < 0.$$

Then, normalized SOP [5.10] weakly converges $v^{\varepsilon}(t) \to \varsigma(t)\ \forall (t_0, t] \subset (0, T]$ at $\varepsilon \to 0,\ T > 0$, where $\varsigma(t), t \geq 0$, is an Ornstein–Uhlenbeck process with the generator

$$L\varphi(v) = bv\varphi'(v) + \frac{\sigma^2}{2}\varphi''(v)$$

and the variance $\sigma^2 = a^2\rho^2$.

CONCLUSION 5.1.– The Ornstein–Uhlenbeck limit process $\varsigma(t)$ is ergodic with stationary normal distribution $N(0, \sigma_0^2)$,

$$\sigma_0^2 = -\frac{\sigma^2}{2b}.$$

CONCLUSION 5.2.– Condition $C1$ defines the diffusivity of the process $\zeta(t)$, and condition $C3$ defines ergodicity.

Proof of Theorem 5.3 is based on the following lemmas.

LEMMA 5.3.– *Normalized SOP [5.10] is a solution of the SDE:*

$$\frac{dv^{\varepsilon}(t)}{dt} = \varepsilon^{-1}\frac{a}{\sqrt{t}}\nabla_{b(t)}C(\frac{\varepsilon v}{\sqrt{t}}, x) + \frac{1}{2t}v. \qquad [5.11]$$

PROOF.– Since

$$v^{\varepsilon}(t) = \sqrt{t}u^{\varepsilon}(t)/\varepsilon,$$

then

$$u^\varepsilon(t) = \varepsilon v^\varepsilon(t) t^{-1/2}.$$

Taking the derivative in the last equality at t, we obtain

$$\frac{du^\varepsilon(t)}{dt} = \varepsilon(-\frac{1}{2})t^{-3/2}v^\varepsilon(t) + \varepsilon t^{-1/2}\frac{dv^\varepsilon(t)}{dt}. \qquad [5.12]$$

Then, we equate the right-hand-side of equalities [5.12] and [5.8]:

$$\varepsilon(-\frac{1}{2})t^{-3/2}v^\varepsilon(t) + \varepsilon t^{-1/2}\frac{dv^\varepsilon(t)}{dt} = \frac{a}{t}\nabla_{b(t)}C(\varepsilon v^\varepsilon(t)/\sqrt{t}, x_t^\varepsilon). \qquad [5.13]$$

Doing elementary transformations in [5.13], we have

$$\begin{aligned}\frac{dv^\varepsilon(t)}{dt} &= \varepsilon^{-1}\sqrt{t}\left[\frac{a}{t}\nabla_{b(t)}C(\varepsilon v^\varepsilon(t)/\sqrt{t}, x_t^\varepsilon) + \frac{1}{2}\varepsilon t^{-3/2}v^\varepsilon(t)\right] \\ &= \frac{a}{\varepsilon\sqrt{t}}\nabla_{b(t)}C(\frac{\varepsilon v^\varepsilon(t)}{\sqrt{t}}, x_t^\varepsilon) + \frac{1}{2t}v^\varepsilon(t).\end{aligned}$$

Consequently, we obtain [5.11].

LEMMA 5.4.– *The generator $\boldsymbol{L}_t^\varepsilon$ of the two-component Markov process $v^\varepsilon(t)$, $x_t^\varepsilon := x(t/\varepsilon^2), t \geq 0$, on the Banach space $B(R, X)$ of real-valued functions $\varphi(v, x) \in C^2(R^d, X)$, is defined by the relation:*

$$\boldsymbol{L}_t^\varepsilon\varphi(v, x) = \varepsilon^{-2}Q\varphi(v, x) + \boldsymbol{C}_t(x)\varphi(v, x), \qquad [5.14]$$

where

$$\boldsymbol{C}_t(x)\varphi(v, x) = [\varepsilon^{-1}\frac{a}{\sqrt{t}}\nabla_{b(t)}C(\frac{\varepsilon v}{\sqrt{t}}, x) + \frac{1}{2t}v]\varphi'(v, x). \qquad [5.15]$$

PROOF.– By the definition of the generated operator of the Markov process, we have the relation

$$\begin{aligned}\mathbf{L}_t^\varepsilon\varphi(v, x) &= \lim_{\Delta\to 0}\Delta^{-1}[E_{v,x}\varphi(v^\varepsilon + \Delta v^\varepsilon, x_{t+\Delta}^\varepsilon) \\ &\quad - \varphi(v, x)|v^\varepsilon(t) = v, x_{t+\Delta}^\varepsilon = x] \\ &= \lim_{\Delta\to 0}\Delta^{-1}E_{v,x}[\varphi(v + \Delta v, x_{t+\Delta}^\varepsilon) - \varphi(v, x)].\end{aligned}$$

For the first summand, we obtain decomposition

$$E_{v,x}[\varphi(v + \Delta v, x_{t+\Delta})] = E_{v,x}\varphi(v, x_{t+\Delta})$$

$$+E_{v,x}[\varphi(v + \Delta v, x_{t+\Delta}) - \varphi(v, x_{t+\Delta})].$$

Considering the time distribution θ_x of staying MP $x(t)$,$t \geq 0$, at state x, we have

$$E_{v,x}[\varphi(v+\Delta v, x_{t+\Delta})] = E_{v,x}\varphi(v, x_{t+\Delta}) + E_{v,x}[\varphi(v+\Delta v, x)$$
$$-\varphi(v,x)]I(\theta_x > \varepsilon^{-2}\Delta) + E_{v,x}[\varphi(v+\Delta v, x_{t+\Delta})$$
$$-\varphi(v, x_{t+\Delta})]I(\theta_x \leq \varepsilon^{-2}\Delta).$$

For the continuous function $\varphi(v,x)$ at v, we obtain:

$$E_{v,x}[\varphi(v+\Delta v, x_{t+\Delta}) - \varphi(v, x_{t+\Delta})]I(\theta_x \leq \varepsilon^{-2}\Delta) = o(\Delta).$$

Similarly,

$$E_{v,x}[\varphi(v+\Delta v, x) - \varphi(v,x)]I(\theta_x > \varepsilon^{-2}\Delta)$$
$$= E_{v,x}[\varphi'(v,x)[\varepsilon^{-1}\frac{a}{\sqrt{t}}\nabla_b C(\frac{\varepsilon v}{\sqrt{t}}, x) + \frac{1}{2t}v]\Delta$$
$$-\varepsilon^{-2}q(x)\varphi'(v,x)\frac{a}{\varepsilon\sqrt{t}}\nabla_b C(\frac{\varepsilon v}{\sqrt{t}}, x)\Delta^2 + o(\Delta)].$$

Therefore, the generator has the form

$$\lim_{\Delta\to 0}\Delta^{-1}\{E_{v,x}[\varphi(v, x_{t+\Delta}) - \varphi(v,x)]$$
$$+E_{v,x}[\varphi(v+\Delta v, x_{t+\Delta}) - \varphi(v, x_{t+\Delta})]\}$$
$$= \varepsilon^{-2}Q\varphi(v,x) + [\varepsilon^{-1}\frac{a}{\sqrt{t}}\nabla_{b(t)}C(\frac{\varepsilon v}{\sqrt{t}}, x) + \frac{1}{2t}v]\varphi'(v,x),$$

whence

$$E_{v,x}[\varphi(v, x_{t+\Delta}) - \varphi(v,x)]$$
$$= E_{v,x}[\varphi(v, x_{t+\Delta}) - \varphi(v,x)][I(\theta_x > \varepsilon^{-2}\Delta) + I(\theta_x \leq \varepsilon^{-2}\Delta]$$
$$= E_{v,x}[\varphi(v, x_{t+\Delta}) - \varphi(v,x)][\varepsilon^{-2}q(x)\Delta + o(\Delta)]$$
$$= \varepsilon^{-2}q(x)E_{(v,x)}[\varphi(v, x_{t+\Delta}) - \varphi(v,x)]\Delta + o(\Delta).$$

LEMMA 5.5.– *The generator* $\boldsymbol{L}_t^\varepsilon$ *has the analytic representation*

$$L_t^\varepsilon\varphi(v,x) = \varepsilon^{-2}Q\varphi(v,x) + \varepsilon^{-1}\frac{1}{\sqrt{t}}Q_1(x)\varphi(v,x)$$
$$+\frac{1}{t}Q_2(x)\varphi(v,x) + o(\varepsilon^2) \qquad [5.16]$$

on the test functions $\varphi(v,\cdot) \in C^2(R)$ *for* $C(u,\cdot) \in C^3(R)$, *where*

$$Q_1(x)\varphi(v,x) = aC'(0,x)\varphi'(v,x),$$

$$Q_2(x)\varphi(v,x) = v(aC''(0,x) + \frac{1}{2})\varphi'(v,x).$$

PROOF.– We consider the function $\nabla_{b(t)}C(\frac{\varepsilon v}{\sqrt{t}},x)$. Since $C(u,\cdot)\in C^3(R)$, it can be decomposed into a Taylor series in some neighborhoods of zero:

$$\nabla_{b(t)}C(\frac{\varepsilon v}{\sqrt{t}},x)=C'(0,x)+\varepsilon\frac{v}{\sqrt{t}}C''(0,x)+o(\varepsilon^2). \quad [5.17]$$

Substituting [5.17] into [5.15] and doing transformations, we obtain

$$\begin{aligned}\mathbf{C}_t(x)\varphi(v,x)&=[\varepsilon^{-1}\frac{a}{\sqrt{t}}[C'(0,x)+\varepsilon\frac{v}{\sqrt{t}}C''(0,x)]+\frac{1}{2t}v]\varphi'(v,x)+o(\varepsilon^2)\\&=[\varepsilon^{-1}\frac{a}{\sqrt{t}}C'(0,x)+\frac{av}{\sqrt{t}}C''(0,x)+\frac{1}{2t}v]\varphi'(v,x)+o(\varepsilon^2).\end{aligned}$$

Consequently, the generator [5.14] considering the last relation, $Q_1(x)$, $Q_2(x)$, we have [5.16].

Furthermore, we consider the test function in the form

$$\varphi^\varepsilon(v,x)=\varphi(v)+\frac{\varepsilon}{\sqrt{t}}\varphi_1(v,x)+\frac{\varepsilon^2}{t}\varphi_2(v,x),\varphi(v)\in C^3(R).$$

LEMMA 5.6.– *The solution of the singular perturbation problem for the generator* $\boldsymbol{L}_t^\varepsilon$ *on the test functions* $\varphi^\varepsilon(v,x)$ *has the form:*

$$\boldsymbol{L}_t^\varepsilon\varphi^\varepsilon(v,x)=\frac{1}{t}\boldsymbol{L}_t\varphi(v)+\varepsilon\theta_t^\varepsilon(x)\varphi(v),$$

where $\boldsymbol{L}_t$ *is a limit generator in the form:*

$$\boldsymbol{L}_t\varphi(v)=v(aC_2+\frac{1}{2})\varphi'(v)+\frac{a^2\rho^2}{2}\varphi''(v),$$

$$C_2=\int_X\pi(dx)C''(0,x),\rho^2=2\int_X\pi(dx)C'(0,x)R_0C'(0,x),$$

and the remainder term is bounded:

$$\|\theta_t^\varepsilon(x)\varphi(v)\|\le M<\infty,$$

which has the representation

$$\theta_t^\varepsilon(x)\varphi(v)=\frac{1}{t\sqrt{t}}[Q_1(x)\varphi_2(v,x)+Q_2(x)\varphi_1(v,x)+\frac{\varepsilon}{\sqrt{t}}Q_2(x)\varphi_2(v,x)].$$

PROOF.– We consider the action of the truncated generator

$$\hat{\boldsymbol{L}}_t^\varepsilon\varphi^\varepsilon=\varepsilon^{-2}Q+\frac{\varepsilon^{-1}}{\sqrt{t}}Q_1(x)+\frac{1}{t}Q_2(x)$$

on the test functions $\varphi^\varepsilon(v,x)$.

Therefore, we obtain

$$\hat{\boldsymbol{L}}_t^{\varepsilon}\varphi^{\varepsilon}(v,x) = \varepsilon^{-2}Q\varphi(v) + \frac{\varepsilon^{-1}}{\sqrt{t}}[Q\varphi_1(v,x) + Q_1(x)\varphi(v)]$$
$$+\frac{1}{t}[Q\varphi_2(v,x) + Q_1(x)\varphi_1(v,x) + Q_2(x)\varphi/(v)]$$
$$+\frac{\varepsilon}{t\sqrt{t}}[Q_1(x)\varphi_2(v,x) + Q_2(x)\varphi_1(v,x) + \frac{\varepsilon}{\sqrt{t}}Q_2(x)\varphi_2(v,x)]. \quad [5.18]$$

Since the operator Q acts only at the variable x, for the first addition in [1.19], we have $Q\varphi \equiv 0$.

Furthermore, we use the balance condition

$$\int_X \pi(dx)Q_1(x) = 0,$$

because $\int_X \pi(dx)C'(0,x) = C'(0) = 0$, and the additional condition $\Pi\varphi(v) = 0$. Then, for the second summand in [5.18], we have:

$$Q\varphi_1(v,x) + Q_1(x)\varphi(v) = 0.$$

From here, we obtain the form of the function $\varphi_1(v,x)$:

$$\varphi_1(v,x) = R_0 Q_1(x)\varphi(v) = aR_0 C'(0,x)\varphi'(v).$$

Furthermore, we use the second solvability condition of singular perturbation and the additional condition $\Pi\varphi(v) = 0$.

For the third summand, we have

$$Q\varphi_2(v,x) + \mathbf{L}_t(x)\varphi(v) = \mathbf{L}_t\varphi(v), \quad [5.19]$$

where

$$\mathbf{L}_t(x)\varphi(v) = Q_1(x)\varphi_1(v,x) + Q_2(x)\varphi(v),$$

and $\mathbf{L}_t = \Pi\mathbf{L}_t(x)$.

From [5.19], we obtain the representation for $\varphi_2(v,x)$

$$\varphi_2(v,x) = R_0\tilde{\mathbf{L}}_t\varphi(v), \quad [5.20]$$

where

$$\tilde{\boldsymbol{L}}_t = \boldsymbol{L}_t(x) - \boldsymbol{L}_t.$$

From the representation $\varphi_1(v,x)$, $Q_1(x)$ and $Q_2(x)$, we have

$$\boldsymbol{L}_t(x)\varphi(v) = a^2C'(0,x)R_0C'(0,x)\varphi''_v(v) + v(aC''(0,x)+\frac{1}{2})\varphi'_v(v).$$

From the last equality and representation $\varphi_2(v,x)$, we obtain the statement of Lemma 5.6.

PROOF OF THEOREM 5.3.–

Using the statement of Lemma 5.6, we can easily check the performance of the conditions of the model limit theorem, i.e. the weak convergence $v^\varepsilon(t) \to \zeta(t)$ at $\varepsilon \to 0$ holds true.

5.1.3. *SOP with impulse perturbation*

One of the problems of system analysis under the uncertainty described by the Markov process is the establishment of convergence conditions of the stochastic optimization procedure (SOP) by the direct influence of the regression function from such process.

The difference stochastic optimization procedure in an ergodic Markov environment with impulse perturbation is given by the evolutionary equation

$$du^\varepsilon(t) = a(t)[\nabla_{b(t)}C(u^\varepsilon(t), x(t/\varepsilon^2))dt + \varepsilon d\eta^\varepsilon(t)], \qquad [5.21]$$

where $C(u,\cdot) \in C^3(R)$.

The Markov process $x(t), t \geq 0$ in a standard phase space $(X, \boldsymbol{X})$ is given by the generator:

$$\mathbf{Q}\varphi(x) = q(x)\int_X P(x,dy)\left[\varphi(y)-\varphi(x)\right], \quad \varphi \in \mathbf{B}(X),$$

where $\mathbf{B}(X)$ is the Banach space of real-valued bounded functions with the norm

$$||\varphi|| = \max_{x\in X}|\varphi(x)|.$$

The stochastic kernel $P(x,B), x \in X, B \in \boldsymbol{X}$ defines a uniformly ergodic embedded Markov chain $x_n = x(\tau_n)$, $n \geq 0$, with the stationary distribution $\rho(B), B \in \boldsymbol{X}$. The stationary distribution $\pi(B), B \in \boldsymbol{X}$, of the Markov process $x(t), t \geq 0$, is defined by the relation:

$$\pi(dx)q(x) = q\rho(dx),\ q = \int_X \pi(dx)q(x).$$

The potential operator R_0 of the generator $\mathbf{Q}$ is defined by the relation: $R_0 = \Pi - (\Pi + \mathbf{Q})^{-1}$, where $\Pi\varphi(x) = \int\limits_X \pi(dy)\varphi(y)$ is projector onto the subspace $N_Q = \{\varphi : \mathbf{Q}\varphi = 0\}$ of zeros of the operator $\mathbf{Q}$.

The impulse perturbation process (IPP) $\eta^\varepsilon(t), t \geq 0$, is given by the relation:

$$\eta^\varepsilon(t) = \int\limits_0^t \eta^\varepsilon(ds; x(s/\varepsilon^2));$$

where the family of processes with independent increments $\eta^\varepsilon(t, x), t \geq 0, x \in X$, is given by the generators

$$\mathbf{\Gamma}^\varepsilon_w(x)\varphi(w) = \varepsilon^{-2} \int\limits_R [\varphi(w + \varepsilon v) - \varphi(w)]\Gamma(dv; x), x \in X. \quad [5.22]$$

We assume that the following balance condition is fulfilled:

$$\Pi\mathbf{\Gamma_1}(x) = \int\limits_X \pi(dx)b_1(x) = 0, b_1(x) = \int\limits_R v\Gamma(dv; x). \quad [5.23]$$

We also consider the generator $\mathbf{\Gamma}^\varepsilon_u(x)$ similarly acting on the first parameter of the function $\varphi(u, w, x)$ [5.22] (at the same time, $\mathbf{\Gamma}^\varepsilon_w(x)$ acts on the second parameter). Therefore

$$\mathbf{\Gamma}^\varepsilon_u(x)\varphi(u, w, x) = \varepsilon^{-2} \int\limits_R [\varphi(u + \varepsilon v, w, x) - \varphi(u, w, x)]\Gamma(dv; x), \quad [5.24]$$

$x \in X$.

The average system is considered in the next representation:

$$\frac{du}{dt} = C'(u), C(u) = \int\limits_X \pi(dx)C(u; x). \quad [5.25]$$

We also introduce the operators

$$\mathbf{\Gamma_1}(x)\varphi(w) = b_1(x)\varphi'(w); \mathbf{\Gamma_2}(x)\varphi(w) = \frac{1}{2}b_2(x)\varphi''(w);$$

$$b_2(x) = \int\limits_R v^2\Gamma(dv; x).$$

LEMMA 5.7.– *Generators of family of processes with independent increments* $\eta^\varepsilon(t,x)$, $t \geq 0$, $x \in X$ *on the test functions* $\varphi(w) \in C^3(R)$ *have the asymptotic representation:*

$$\mathbf{\Gamma}^\varepsilon_w(x)\varphi(w) = \varepsilon^{-1}\mathbf{\Gamma_1}(x)\varphi(w) + \mathbf{\Gamma_2}(x)\varphi(w) + \gamma^\varepsilon(x)\varphi(w), \qquad [5.26]$$

where the remainder term is such that $||\gamma^\varepsilon(x)\varphi(w)|| \to 0$ *at* $\varepsilon \to 0$*;* $\varphi(w) \in C^3(R)$.

PROOF.– Using the decomposition of $varphi(w)$ into a Taylor series, we transform the generator

$$\begin{aligned}\mathbf{\Gamma}^\varepsilon_w(x)\varphi(w) &= \varepsilon^{-2}\int_R [\varphi(w+\varepsilon v) - \varphi(w)]\Gamma(dv;x)\\ &= \varepsilon^{-2}\int_R [\varphi(w+\varepsilon v) - \varphi(w) - \varepsilon v\varphi'(w) - \frac{1}{2}\varepsilon^2 v^2\varphi''(w)]\Gamma(dv;x)\\ &\quad + \varepsilon^{-1}b_1(x)\varphi'(w) + \frac{1}{2}b_2(x)\varphi''(w)\\ &= \varepsilon^{-1}b_1(x)\varphi'(w) + \frac{1}{2}b_2(x)\varphi''(w) + \gamma^\varepsilon(x)\varphi(w).\end{aligned}$$

Considering that $\gamma^\varepsilon(x)\varphi(w) = O(\varepsilon)$, $\varphi(w) \in C^3(R)$, we obtain the required representation.

THEOREM 5.4.– *Let there exist a Lyapunov function* $V(u) \in C^3(R)$ *such that the exponential stability of the average system [5.25] holds true:*

$$C1 : C'(u)V'(u) \leq -cV(u), c > 0, \qquad [5.27]$$

and the additional conditions

$$\begin{aligned}&C2 : |\mathbf{\Gamma_2}(x)V(u)| \leq c_2(1+V(u)), c_2 > 0,\\ &C3 : |\mathbf{\Gamma_1}(x)R_0\mathbf{\Gamma_1}(x)V(u)| \leq c_4(1+V(u)), c_4 > 0,\\ &C4 : |\mathbf{\Gamma_1}(x)R_0\tilde{\mathbf{C}}(x)V(u)| \leq c_5(1+V(u)), c_5 > 0,\\ &C5 : |\mathbf{C}(x)R_0\mathbf{\Gamma_1}(x)V(u)| \leq c_6(1+V(u)), c_6 > 0,\\ &C6 : |\mathbf{C}(x)R_0\tilde{\mathbf{C}}(x)V(u)| \leq c_7(1+V(u)), c_7 > 0, \qquad [5.28]\\ &C7 : |\mathbf{\Gamma_2}(x)R_0\mathbf{\Gamma_1}(x)V(u)| \leq c_8(1+V(u)), c_8 > 0,\\ &C8 : |\mathbf{\Gamma_2}(x)R_0\tilde{\mathbf{C}}(x)V(u)| \leq c_9(1+V(u)), c_9 > 0,\\ &C9 :| (\nabla_{b(t)}C(u) - C'(u))V'(u) |\leq c_1 b(t)(1+V(u)),\end{aligned}$$

where

$$\tilde{\mathbf{C}}(x) = \mathbf{C}(x) - \mathbf{L}.$$

In addition, let the function $C(u,x)$ have two derivatives at $u \in R$ and together with functions $b_1(x)$ and $b_2(x)$ is uniformly bounded at $x \in X$.

We also assume that the balance condition [5.23] is fulfilled.

Let the control function $a(t) > 0$ satisfy the following conditions:

$$\int_0^\infty a(t)dt = \infty, \quad \int_0^\infty a^2(t)dt < \infty, \int_0^\infty a(t)b(t)dt < \infty$$

Then, for each $\varepsilon < \varepsilon_0$, ε_0 is small enough and the solution of the evolutionary equation [5.21] at all initial conditions $u^\varepsilon(0) = u$ with probability 1 converges to the minimum point $u_0 = 0$ of the averaged system [5.22]:

$$P\{\lim_{t\to\infty} u^\varepsilon(t) = 0\} = 1.$$

Furthermore, we prove some of the lemmas.

LEMMA 5.8.– *The generator of a two-component Markov process $\eta^\varepsilon(t)$, $x(t/\varepsilon^2)$, $t \geq 0$, has the form:*

$$\begin{aligned}\hat{\Gamma}^\varepsilon(x)\varphi(w,x) &= \varepsilon^{-2}\mathbf{Q}\varphi(w,x) + \varepsilon^{-1}\mathbf{\Gamma}_1(x)\varphi(w,x) \\ &+ \mathbf{\Gamma}_2(x)\varphi(w,x) + \gamma^\varepsilon(x)\varphi(w,x),\end{aligned} \quad [5.29]$$

where the remainder term $||\gamma^\varepsilon(x)\varphi(w,x)|| \to 0$ at $\varepsilon \to 0$, $\varphi(w,\cdot) \in C^3(R)$.

PROOF.– To prove this statement, we can use the definition of the generator of the Markov process and the forms of the respective generators of processes $\eta^\varepsilon(t)$ and $x(t/\varepsilon^2)$.

The truncated operator has the form:

$$\mathbf{\Gamma}_0^\varepsilon(x)\varphi(w,x) = \varepsilon^{-2}\mathbf{Q}\varphi(w,x) + \varepsilon^{-1}\mathbf{\Gamma}_1(x)\varphi(w,x) + \mathbf{\Gamma}_2(x)\varphi(w,x). \quad [5.30]$$

LEMMA 5.9.– *Under the balance condition [5.23], the solution of the singular perturbation problem for the truncated operator [5.30] on the test functions*

$$\varphi^\varepsilon(w,x) = \varphi(w) + \varepsilon\varphi_1(w,x) + \varepsilon^2\varphi_2(w,x)$$

is determined by the relation:

$$\mathbf{\Gamma}_0^\varepsilon(x)\varphi^\varepsilon(w,x) = \mathbf{\Gamma}\varphi(w) + \varepsilon\theta_\eta^\varepsilon(x)\varphi(w), \quad [5.31]$$

where the remainder term $\theta_\eta^\varepsilon(x)\varphi(w)$ is uniformly bounded at x.

The limit operator $\mathbf{\Gamma}$ *is defined by the formula:*

$$\mathbf{\Gamma}\Pi\varphi(w) = \Pi\mathbf{\Gamma_2}(x)\Pi\varphi(w) + \Pi\mathbf{\Gamma_1}(x)R_0\mathbf{\Gamma_1}(x)\Pi\varphi(w), \qquad [5.32]$$

and the remainder term $\theta_\eta^\varepsilon(x)\varphi(w) = \mathbf{\Gamma_1}(x)R_0\tilde{\mathbf{\Gamma}}(x)\varphi(w) + \mathbf{\Gamma_2}(x)R_0\mathbf{\Gamma_1}(x)\varphi(w) + \varepsilon\mathbf{\Gamma_2}(x)R_0\tilde{\mathbf{\Gamma}}(x)\varphi(w)$.

PROOF.– For equality [5.31], it is necessary that the coefficients of equal degrees of ε on the left and the right coincide.

Let us calculate

$$\begin{aligned}\mathbf{\Gamma}_0^\varepsilon(x)\varphi^\varepsilon(w,x) &= \varepsilon^{-2}\mathbf{Q}\varphi(w) + \varepsilon^{-1}[\mathbf{Q}\varphi_1(w,x) + \mathbf{\Gamma_1}(x)\varphi(w)] \\ &+ [\mathbf{Q}\varphi_2(w,x) + \mathbf{\Gamma_1}(x)\varphi_1(w,x) + \mathbf{\Gamma_2}(x)\varphi(w)] \\ &+ \varepsilon[\mathbf{\Gamma_1}(x)\varphi_2(w,x) + \mathbf{\Gamma_2}(x)\varphi_1(w,x) + \varepsilon\mathbf{\Gamma_2}(x)\varphi_2(w,x)].\end{aligned}$$

Since $\varphi(w)$ does not depend on x,

$$\mathbf{Q}\varphi(w) = 0, \Leftrightarrow \varphi(w) \in N_{\mathbf{Q}}.$$

The balance condition [5.23] is a solvability condition for the equation

$$\mathbf{Q}\varphi_1(w,x) + \mathbf{\Gamma_1}(x)\varphi(w) = 0.$$

Consequently,

$$\varphi_1(w,x) = R_0\mathbf{\Gamma_1}(x)\varphi(w). \qquad [5.33]$$

Using [5.33], the equation

$$\mathbf{Q}\varphi_2(w,x) + \mathbf{\Gamma_1}(x)\varphi_1(w,x) + \mathbf{\Gamma_2}(x)\varphi(w) = \mathbf{\Gamma}\varphi(w)$$

can be rewritten as

$$\mathbf{Q}\varphi_2(w,x) + \mathbf{\Gamma_1}(x)R_0\mathbf{\Gamma_1}(x)\varphi(w) + \mathbf{\Gamma_2}(x)\varphi(w) = \mathbf{\Gamma}\varphi(w)$$

The solvability condition of the last equation allows us to obtain the limit operator $\mathbf{\Gamma}$ in the form [5.32].

Then

$$\varphi_2(w,x) = R_0[\mathbf{\Gamma_1}(x)R_0\mathbf{\Gamma_1}(x) + \mathbf{\Gamma_2}(x) - \mathbf{\Gamma}]\varphi(w). \qquad [5.34]$$

Using [5.33] and [5.34], we obtain the remainder term in the required form.

Boundedness of $\theta_\eta^\varepsilon(x)\varphi(w)$ follows from the properties of the generators $\mathbf{\Gamma_1}(x)$, $\mathbf{\Gamma_2}(x)$ and R_0.

The theorem is completed using Lemma 5.9 and Korolyuk's limit model theorem.

Now we can consider a two-component Markov process:

$$u^\varepsilon(t),\ x_t^\varepsilon := x(t/\varepsilon^2),\ t \geq 0. \qquad [5.35]$$

Since the SOP $u^\varepsilon(t)$ depends on the control function $a(t)$, this process is not uniform in time.

LEMMA 5.10.– *The generator of the two-component Markov process $u^\varepsilon(t)$, $x(t/\varepsilon^2)$, $t \geq 0$ has the form:*

$$\mathbf{L}_t^\varepsilon(x)\varphi(u,x) = \varepsilon^{-2}\mathbf{Q}\varphi(u,x) + \varepsilon a(t)\mathbf{\Gamma}_u^\varepsilon(x)\varphi(u,x) + a(t)\mathbf{C}^\nabla(x)\varphi(u,x), \qquad [5.36]$$

where $\mathbf{C}^\nabla(x)\varphi(u,x) = \nabla_{b(t)}C(u,x)\varphi_u'(u,x)$,

PROOF.– Let us denote $u_t = u^\varepsilon(t)$, $x_t = x(t/\varepsilon^2)$. Then, using the definition of the generator of the Markov process, we have

$$\begin{aligned}\mathbf{L}_t^\varepsilon(x)\varphi(u,x) &= \lim_{\Delta\to 0}\frac{1}{\Delta}E[\varphi(u^\varepsilon(t+\Delta), x((t+\Delta)/\varepsilon^2))\\ &\quad - \varphi(u^\varepsilon(t), x(t/\varepsilon^2))|u^\varepsilon(t) = u, x(t/\varepsilon^2) = x]\\ &= \lim_{\Delta\to 0}\frac{1}{\Delta}E[\varphi(u_{t+\Delta}, x_{t+\Delta}) - \varphi(u,x)]\\ &= \lim_{\Delta\to 0}\frac{1}{\Delta}E[\varphi(u_{t+\Delta}, x_{t+\Delta}) - \varphi(u, x_{t+\Delta})]\\ &\quad + \lim_{\Delta\to 0}\frac{1}{\Delta}E[\varphi(u, x_{t+\Delta}) - \varphi(u,x)]\end{aligned} \qquad [5.37]$$

Since the representation

$$u(t+\Delta) = u + a(t)\nabla_b(t)C(u,x)\Delta + \varepsilon a(t)\Delta\eta^\varepsilon(t) + o(\Delta)$$

holds true, for the first summand in [5.37], we have

$$\lim_{\Delta\to 0}\frac{1}{\Delta}E[\varphi(u_{t+\Delta}, x_{t+\Delta}) - \varphi(u, x_{t+\Delta})]$$

$$\begin{aligned}&= \lim_{\Delta\to 0}\frac{1}{\Delta}E[\varphi(u + a(t)\nabla_b(t)C(u,x)\Delta\\ &\quad +\varepsilon a(t)\Delta\eta^\varepsilon(t) + o(\Delta), x_{t+\Delta}) - \varphi(u, x_{t+\Delta})].\end{aligned}$$

Using summand

$$\pm\varphi(u + a(t)\nabla_b(t)C(u,x)\Delta + o(\Delta), x_{t+\Delta}),$$

we obtain:

$$\lim_{\Delta\to 0}\frac{1}{\Delta}E[\varphi(u_{t+\Delta}, x_{t+\Delta}) - \varphi(u, x_{t+\Delta})]$$

$$= \lim_{\Delta\to 0}\frac{1}{\Delta}E[\varphi(u + a(t)\nabla_b(t)C(u,x)\Delta + \varepsilon a(t)\Delta\eta^\varepsilon(t) + o(\Delta), x_{t+\Delta})$$
$$-\varphi(u + a(t)\nabla_b(t)C(u,x)\Delta + o(\Delta), x_{t+\Delta})]$$
$$+\lim_{\Delta\to 0}\frac{1}{\Delta}E[\varphi(u + a(t)\nabla_b C(u,x)\Delta + o(\Delta), x_{t+\Delta}) - \varphi(u, x_{t+\Delta})].$$

From the definition of the generator $\mathbf{\Gamma}^\varepsilon(x)$, we have

$$\mathbf{\Gamma}_u^\varepsilon(x)\varphi(u) = \lim_{\Delta\to 0}\frac{1}{\Delta}E[\varphi(u + \Delta\eta^\varepsilon(t)) - \varphi(u)].$$

If we replace $v = u + a(t)\nabla_b(t)C(u,x)\Delta + o(\Delta)$ (we note that $v \to u\ \Delta \to 0$), we obtain the form for the first limit:

$$\lim_{\Delta\to 0}\frac{1}{\Delta}E[\varphi(v + \varepsilon a(t)\Delta\eta^\varepsilon(t), x_{t+\Delta})$$
$$-\varphi(v, x_{t+\Delta})] = \varepsilon a(t)\mathbf{\Gamma}_u^\varepsilon(x)\varphi(u,x).$$

After transformations, the second limit has the form:

$$\lim_{\Delta\to 0}\frac{1}{\Delta}E[\varphi(u + a(t)\nabla_b(t)C(u,x)\Delta, x_{t+\Delta}) - \varphi(u, x_{t+\Delta})]$$

$$= \lim_{\Delta\to 0}\frac{1}{\Delta}E[\varphi'_u(u, x_{t+\Delta})(a(t)\nabla_b(t)C(u,x)\Delta + o(\Delta))]$$
$$= a(t)\nabla_b(t)C(u,x)\varphi'_u(u,x).$$

Using the definition of the generator of the Markov process $x(t/\varepsilon^2)$:

$$\lim_{\Delta\to 0}\frac{1}{\Delta}E[\varphi(u, x_{t+\Delta}) - \varphi(u,x)] = \varepsilon^{-2}\mathbf{Q}\varphi(u,x),$$

and summing obtained limits, we have [5.36].

LEMMA 5.11.– *The generator $\mathbf{L}_t^\varepsilon(x)$ of the two-component Markov process $u^\varepsilon(t)$, $x(t/\varepsilon^2)$, $t \geq 0$ has the asymptotic representation*

$$\mathbf{L}_t^\varepsilon(x)\varphi(u,x) = \varepsilon^{-2}\mathbf{Q}\varphi(u,x) + a(t)\mathbf{\Gamma_1}(x)\varphi(u,x)$$
$$+ a(t)\mathbf{C}^\nabla(x)\varphi(u,x) + \theta_t^\varepsilon(x)\varphi(u,x), \qquad [5.38]$$

where

$$\theta_t^\varepsilon(x) = \varepsilon a(t)\mathbf{\Gamma_2}(x)\varphi(u,x).$$

The remainder term is such that $||\theta_t^\varepsilon(x)(x)\varphi(u,x)|| \to 0$ at $\varepsilon \to 0$.

PROOF OF THE THEOREM.– We can provide it using the representation of the operator $\mathbf{\Gamma}_u^\varepsilon(x)$ and the results of Lemma 5.10.

LEMMA 5.12.– *The solution of the singular perturbation problem for the operator [1.39] on the perturbed Lyapunov function*

$$V^\varepsilon(u,x) = V(u) + \varepsilon^2 a(t) V_0(u,x)$$

has the representation

$$\mathbf{L}_t^\varepsilon(x) V^\varepsilon(u,x) = a(t)\mathbf{L}V(u) + \varepsilon\theta_t^\varepsilon(x)V(u), \qquad [5.39]$$

where the limit generator $\mathbf{L}$ *has the form*

$$\mathbf{L}V(u) = \nabla_{b(t)}\mathbf{C}(u)V'(u), \nabla_{b(t)}\mathbf{C}(u) = \int_X \pi(dx)\nabla_{b(t)}\mathbf{C}(u,x). \qquad [5.40]$$

and the remainder term $\theta_t^\varepsilon(x)$ *satisfies the inequality*

$$|\theta_t^\varepsilon(x)V(u)| \le c^* a^2(t)(1 + V(u)). \qquad [5.41]$$

PROOF.– The decomposition of the generator [5.38] on the perturbed Lyapunov function has the form

$$\begin{aligned}
&\mathbf{L}_t^\varepsilon(x)V_t^\varepsilon(u,x) = [\varepsilon^{-2}\mathbf{Q} + a(t)\mathbf{\Gamma}_1(x) + a(t)\mathbf{C}^\nabla(x)] \\
&*[V(u) + \varepsilon^2 a(t)V_0(u,x)] = \varepsilon^{-2}\mathbf{Q}V(u) + a(t)[\mathbf{Q}V_0(u,x) \\
&+\mathbf{C}^\nabla(x)V(u) + \mathbf{\Gamma}_1(x)V(u)] + \varepsilon^2 a^2(t)[\mathbf{\Gamma}_1(x)V_0(u,x) \\
&+\mathbf{C}^\nabla(x)V_2(u,x)].
\end{aligned} \qquad [5.42]$$

Since $V(u)$ does not depend on x,

$$\mathbf{Q}V(u) = 0, \Leftrightarrow V(u) \in N_Q.$$

From the solvability of the equation

$$\mathbf{Q}V_0(u,x) + \mathbf{C}^\nabla(x)V(u) + \mathbf{\Gamma}_1(x)V(u) = \mathbf{L}_t V(u),$$

we obtain the limit generator in the form

$$\mathbf{L}_t = \Pi\mathbf{L}_t = \Pi[\mathbf{C}^\nabla(x) + \mathbf{\Gamma}_1(x)].$$

According to the balance condition [5.23] $\Pi\mathbf{\Gamma}_1(x) = 0$, the limit operator has the form

$$\mathbf{L}_t = \Pi\mathbf{C}^\nabla(x).$$

From condition $C1$ of the theorem and the inequality [5.41], we obtain the inequality:

$$\mathbf{L}_t^\varepsilon(x)V_t^\varepsilon(u,x) \le -ca(t)V(u) + c^* a^2(t)(1 + V(u)). \qquad [5.43]$$

The convergence of the stochastic optimization procedure follows from inequality [5.43] and the Nevelson–Khasminsky theorem (Nevelson and Khasminsky 1972).

5.2. SOP under the diffusion approximation scheme

In this section, we consider a continuous stochastic optimization procedure under the diffusion approximating scheme with a regression function that depends on the Markov switching process and satisfies the Lipschitz condition under the global balance on singular perturbation of the regression function.

5.2.1. *Convergence SOP*

Let $C(u, x)$, $u \in R^d$, be a regression function that depends on the impact of the external environment, which is described by a uniformly ergodic Markov process $x(t)$, $t \geq 0$, on a measurable phase space of states $(X, \mathbf{X})$. The generator

$$Q\varphi(x) = q(x) \int_X P(x, dy)[\varphi(y) - \varphi(x)]$$

of the Markov process $x(t), t \geq 0$, is defined on the Banach space $\boldsymbol{B}(X)$ of real-valued continuous functions $\varphi(x)$, $x \in X$, with the supremum norm $\|\varphi(x)\| = \sup_{x \in X} |\varphi(x)|$, where $P(x, B)$, $x \in X$, $B \in \mathbf{X}$, is a stochastic kernel, $q(x) = [g(x)]^{-1}$, $g(x) = E\theta_x$ and $\theta(x)$ is the duration of the Markov process in the state x. We can construct the potential $R_0 = \Pi - (\Pi + Q)^{-1}$, where $\Pi\varphi(x) = \int_X \pi(dx)\varphi(x)$ is a projector onto the zero-subspace for the operator Q: $N_Q = \{\varphi : Q\varphi = 0\}$ and $\pi(B)$, $B \in \mathbf{X}$, is a stationary distribution of the Markov process.

The continuous stochastic optimization procedure for finding the point of maximum for the regression function $C(u, x)$ under the diffusion approximation scheme is given by the stochastic differential equation:

$$\begin{aligned} du^{\varepsilon}(t) = a(t)[\nabla_{b(t)}C(u^{\varepsilon}(t), x(t/\varepsilon^2)) \\ + \varepsilon^{-1}C_0(u^{\varepsilon}(t), x(t/\varepsilon^2))]dt, \end{aligned} \qquad [5.44]$$

where

$$\begin{aligned} &\nabla_{b(t)}C(u, \cdot) = [(C(u_i^+, \cdot) - C(u_i^-, \cdot))/2b(t), i = \overline{1, d}], \\ &u_i^{\pm} = u_i \pm b(t)e_i,\ e_i = \{0, ..., 1, 0, ..., 0\}, \\ &C_0(u, x) = \{C_{0i}(u, x)\ i = \overline{1, d}\}, u \in R^d, x \in \mathbf{X} \end{aligned}$$

Along with the system [5.44], we consider the parametric system

$$du_x(t) = C^{\varepsilon, \nabla_b}(u_x, x)dt, x \in X,$$

where

$$C^{\varepsilon,\nabla_b}(u,x) = \nabla_{b(t)}C(u,x) + \varepsilon^{-1}C_0(u,x),$$

with the generating operator

$$\boldsymbol{C}^{\varepsilon,\nabla_b}\varphi(u,x) = \left[\nabla_{b(t)}\boldsymbol{C}(x) + \varepsilon^{-1}\boldsymbol{C}_0(x)\right]\varphi(u,x),$$

where

$$\nabla_{b(t)}\boldsymbol{C}(x)\varphi(u,x) = \nabla_{b(t)}C(u,x)\varphi'(u,x),$$
$$\boldsymbol{C}_0(x)\varphi(u,x) = C_0(u,x)\varphi'(u,x).$$

The procedure [5.44] is considered under the balance condition on perturbation $C_0(u,x)$:

$$\int_X \pi(dx)C_0(u,x) = 0. \qquad [5.45]$$

For the averaged regression function

$$C(u) = \Pi C(u,x) = \int_X \pi(dx)C(u,x),$$

we consider the dynamical function

$$\frac{du(t)}{dt} = grad\, C(u(t))^T, grad\; C(u) = \left[\frac{\partial C(u)}{\partial u_i}, i = \overline{1,d}\right]. \qquad [5.46]$$

We assume the existing unique point of maximum for the regression function.

THEOREM 5.5.– *Let the averaged regression function $C(u)$ have a continuous partial derivative at $u \in R^d$, satisfying the global Lipschitz condition*

$$B : |\nabla_{b(t)}C(u) - grad\, C(u)| < kb(t), k > 0;$$

the vector function $C_0(u,x)$ has the first and second derivatives, and the vector function $\nabla_{b(t)}C(u,x)$ has the first and second derivatives at $u \in R^d$, uniformly bounded at $x \in X$, and the Lyapunov function $V(u)$ for the averaged system [5.46], that has bounded derivatives to the third order inclusive, which satisfies the following conditions:

C1: exponential stability of the dynamical system [5.46]

$gradC(u)V'(u) < -cV(u), c > 0;$

C2:$|V'(u)| \leq c_1(1 + V(u)), c_1 > 0;$

C3:$|C_0(u,x)R_0[C_0(u,x)V'(u)]'| \leq c_2(1 + V(u)), c_2 > 0;$

C4:$|\nabla_{b(t)}C(u,x)R_0[C_0(u,x)V'(u)]'| \leq c_3(1+V(u)), c_3 > 0;$

C5:$|C_0(u,x)R_0[\nabla_{b(t)}\widetilde{C}(u,x)V'(u)]'| \leq c_4(1+V(u)), c_4 > 0;$

C6:$|\nabla_{b(t)}C(u,x)R_0[\nabla_{b(t)}\widetilde{C}(u,x)V'(u)]'| \leq c_5(1+V(u)), c_5 > 0;$

C7:$|C_0(u,x)R_0[C_0(u,x)R_0[C_0(u,x)V'(u)]']'| \leq c_6(1+V(u), c_6 > 0;)$

C8:$|\nabla_{b(t)}C(u,x)R_0[C_0(u,x)R_0[C_0(u,x)V'(u)]']'| \leq c_7(1+V(u)), c_7 > 0;$

where $\nabla_{b(t)}\widetilde{C}(u,x) = \nabla_{b(t)}C(u,x) - \nabla_{b(t)}C(u)$.

In addition, the balance condition [5.45] is fulfilled, and for the functions $a(t) > 0$ *,*$b(t) > 0$*, we have*

$$\int_0^\infty a(t)dt = \infty, \int_0^\infty a^2(t)dt < \infty, \int_0^\infty a(t)b(t)dt < \infty.$$

Then, for each initial value $u^\varepsilon(0) = u\, ,u \in R^d$*, the solution of the equation [1.45] at* $\varepsilon \leq \varepsilon_0$ *is small enough, converges with probability 1 to the point of extremum* u_0 *and is defined definitely by the equation* $gradC(u_0) = 0$.

To establish convergence conditions for the stochastic optimization procedure, we construct the generator of the stochastic optimization procedure and investigate its asymptotic representation.

We now consider the two-component Markov process

$$u^\varepsilon(t), x_t^\varepsilon := x(t/\varepsilon^2), t \geq 0. \qquad [5.47]$$

LEMMA 5.13.– *The generator of the two-component Markov process [5.47] on the test functions* $\varphi(u,x) \in C^{2,0}(R^d, X)$ *has the analytical representation*

$$\mathbf{L}_t^\varepsilon(x)\varphi(u,x) = \varepsilon^{-2}\boldsymbol{Q}\varphi(u,x) + \varepsilon^{-1}a(t)\mathbf{C}_0(x)\varphi(u,x) + a(t)\nabla_{b(t)}\mathbf{C}(x)\varphi(u,x).$$

PROOF.– We consider the conditional expectation:

$$E[\varphi(u^\varepsilon(t+\Delta), x_{t+\Delta}^\varepsilon)|u^\varepsilon(t) = u, x_t^\varepsilon = x] = E_{u,x}[\varphi(u^\varepsilon(t+\Delta), x_{t+\Delta}^\varepsilon)]$$

$$= E_{u,x}[\varphi(u + \int_t^{t+\Delta} a(s)C^{\varepsilon,\nabla_b}(u^\varepsilon(s), x(s/\varepsilon^2))\, ds, x]I(\theta_x > \varepsilon^{-2}\Delta)$$

$$+ E_{u,x}\varphi[u + \int_t^{t+\Delta} a(s)C^{\varepsilon,\nabla_b}(u^\varepsilon(s), x(s/\varepsilon^2))\, ds, x_{t+\Delta}]$$

$$\times I(\theta_x \leq \varepsilon^{-2}\Delta) + o(\Delta) \qquad [5.48]$$

The staying time θ_x at state x has an exponential distribution, i.e. we have decompositions:

$$I(\theta_x > \varepsilon^{-2}) = e^{-\varepsilon^{-2}q(x)\Delta} = 1 - \varepsilon^{-2}q(x)\Delta + o(\Delta),$$

$$I(\theta_x \leq \varepsilon^{-2}) = 1 - e^{-\varepsilon^{-2}q(x)\Delta} = \varepsilon^{-2}q(x)\Delta + o(\Delta), \qquad [5.49]$$

where $q(x)$ is the intensivity.

Furthermore, we use transformations for the test functions $\varphi(u, x)$:

$$\varphi[u + \int_t^{t+\Delta} a(s)C^{\varepsilon,\nabla_b}(u^\varepsilon(s), x(s/\varepsilon^2))\, ds, x]$$

$$= \varphi(u, x) + \varphi'_u(u, x) \int_{t+\Delta}^{t} a(s)C^{\varepsilon,\nabla_b}(u^\varepsilon(s), x(s/\varepsilon^2))ds$$

$$= \varphi(u, x) + C^{\varepsilon,\nabla_b}(u, x)\varphi'_u(u, x)a(t)\Delta + o(\Delta) \qquad [5.50]$$

Considering [5.49] and [5.50], we calculate the conditional expectation [5.48]:

$$E_{u,x}[\varphi(u, x) + a(t)C^{\varepsilon,\nabla_b}(u, x)\varphi'_u(u, x)\Delta][1 - \varepsilon^{-2}q(x)\Delta + o(\Delta)]$$
$$+E_{u,x}[\varphi(u, x^\varepsilon_{t+\Delta}) + a(t)C^{\varepsilon,\nabla_b}(u, x)\varphi'_u(u, x^\varepsilon_{t++\Delta})\Delta]$$
$$\times[\varepsilon^{-2}q(x)\Delta + o(\Delta)] = \varphi(u, x) + a(t)C^{\varepsilon,\nabla_b}(u, x)\varphi'_u(u, x)\Delta$$
$$-\varepsilon^{-2}q(x)\Delta[E_{u,x}\varphi(u, x) + o(\Delta)] + \varepsilon^{-2}q(x)\Delta[E_{u,x}\varphi(u, x^\varepsilon_{t+\Delta}) + +o(\Delta)]$$
$$= \varphi(u, x) + a(t)C^{\varepsilon,\nabla_b}(u, x)\varphi'_u(u, x)\Delta+$$
$$+\varepsilon^{-2}q(x)\Delta \int_X P(x, dy)[\varphi(u, y) - \varphi(u, x)] + o(\Delta). \qquad [5.51]$$

Considering [5.51] and the definition of the generator of the two-component Markov process, we obtain

$$\mathbf{L}^\varepsilon_t(x)\varphi(u, x) = \lim_{\Delta \to 0} \frac{1}{\Delta}(E(\varphi(u^\varepsilon(t + +\Delta), x^\varepsilon_{t+\Delta})|u^\varepsilon(t) = u, x^\varepsilon_t = x)$$

$$-\varphi(u, x)) == \varepsilon^{-2}Q\varphi(u, x) + \varepsilon^{-1}a(t)\mathbf{C}_0(x)\varphi(u, x) + a(t)\nabla_{b(t)}\mathbf{C}(x)\varphi(u, x).$$

and the lemma is proved.

Furthermore, we use the perturbed Lyapunov function in the form:

$$V^\varepsilon(u, x) = V(u) + \varepsilon a(t)V_1(u, x) + \varepsilon^2 V_2(u, x),$$

where $V(u) \in C^3(R^d)$ is the Lyapunov function of the averaged system [1.47].

LEMMA 5.14.– *The generator $\mathbf{L}_t^\varepsilon(x)$ on the perturbed Lyapunov function $V^\varepsilon(u,x)$ has the asymptotic decomposition:*

$$\mathbf{L}_t^\varepsilon(x)V^\varepsilon(u,x) = L_tV(u) + \varepsilon\Theta_t(x)V(u),$$

$$\mathbf{L}_tV(u) = [a(t)\Pi\mathbf{C}_0(x)R_0\mathbf{C}_0(x) + \Pi\nabla_{b(t)}\mathbf{C}(x)]V(u),$$

$$\Theta_t(x)V(u) = \mathbf{C}_0(x)R_0\widetilde{\mathbf{L}}_t(x)V(u) + \nabla_b\mathbf{C}(x)R_0\mathbf{C}_0(x)V(u)$$
$$+\varepsilon\nabla_{b(t)}\mathbf{C}_0(x)R_0\widetilde{\mathbf{L}}_t(x)V(u),$$

where

$$\widetilde{\mathbf{L}}_t(x) = \mathbf{L}_t(x) - \mathbf{L}.$$

PROOF.–

The generator on the perturbed Lyapunov function has the representation

$$\mathbf{L}_t^\varepsilon V^\varepsilon(u,x) = \varepsilon^{-2}QV(u) + \varepsilon^{-1}a(t)[a(t)QV_1(u,x) + \mathbf{C}_0(x)V(u)]$$
$$+a(t)[QV_2(u,x) + a(t)\mathbf{C}_0(x)V_1(u,x) + \nabla_{b(t)}\mathbf{C}(x)V(u)]$$
$$+\varepsilon a^2(t)[\mathbf{C}_0(x)V_2(u,x) + \nabla_b\mathbf{C}(x)V_1(u,x) + +\varepsilon\nabla_{b(t)}\mathbf{C}(x)V_2(u,x)].$$

We consider additions grouping by the order of the parameter ε.

Since the function $V(u)$ does not depend on the variable x, the first summand is equal to zero:

$$QV(u) = 0.$$

Considering the first condition of solvability of the singular perturbation problem, we equate the addition with the coefficient ε^{-1} to zero:

$$a(t)QV_1(u,x) + a(t)\mathbf{C}_0(x)V(u) = 0. \quad [5.52]$$

Using the balance condition for equation [5.52] and the additional condition $\Pi V_1 = 0$, we obtain the solution in the form

$$V_1(u,x) = R_0\mathbf{C}_0(x)V(u). \quad [5.53]$$

By the solvability of the singular perturbation problem for the definition of the function $V_2(u,x)$, we have the equality

$$QV_2(u,x) + \mathbf{L}_t(x)V(u) = \mathbf{L}_tV(u),$$

where

$$\boldsymbol{L}_t(x)V(u) = a(t)\boldsymbol{C}_0(x)R_0\boldsymbol{C}_0(x)V(u) + \nabla_{b(t)}\boldsymbol{C}(x)V(u),$$

and $\boldsymbol{L}_t = \Pi\boldsymbol{L}_t(x)$, i.e.

$$\boldsymbol{L}_t = a(t)\Pi\boldsymbol{C}_0(x)R_0\boldsymbol{C}_0(x) + \Pi\nabla_b\boldsymbol{C}(x). \qquad [5.54]$$

Therefore, we obtain

$$V_2(u,x) = R_0\widetilde{\boldsymbol{L}}_t(x)V(u), \qquad [5.55]$$

where $\widetilde{\boldsymbol{L}}_t(x) = \boldsymbol{L}_t(x) - \boldsymbol{L}_t$.

On the fourth summand, we substitute the value of the functions $V_1(u,x)$ [5.53] and $V_2(u,x)$ [5.55], and then mark the remainder term $\Theta_t(x)V(u)$:

$$\begin{aligned}\Theta_t(x)V(u) &= \boldsymbol{C}_0(x)V_2(u,x) + \nabla_{b(t)}\boldsymbol{C}(x)V_1(u,x)\\ &\quad + \varepsilon\nabla_b\boldsymbol{C}(x)V_2(u,x)\\ &= \boldsymbol{C}_0(x)R_0\widetilde{\boldsymbol{L}}_t(x)V(u) + \nabla_{b(t)}\boldsymbol{C}(x)R_0\boldsymbol{C}_0(x)V(u)\\ &\quad + \varepsilon\nabla_{b(t)}\boldsymbol{C}(x)R_0\widetilde{\boldsymbol{L}}_t(x)V(u). \qquad [5.56]\end{aligned}$$

The lemma is proved.

PROOF OF THE THEOREM.–

We consider the estimates for the asymptotical decomposition of the generator $\boldsymbol{L}_t^\varepsilon$.

For example, considering [5.54], conditions $C1 - C3$ and B of Theorem 5.5, for the first summand, we have

$$\begin{aligned}\boldsymbol{L}_tV(u) &= \int_X \pi(dx)\nabla_{b(t)}C(u,x)V'(u)\\ &\quad + a(t)\int_X \pi(x)C_0(u,x)R_0[C_0(u,x)V'(u)]\\ &\le -cV(u) + k^*b(t)(1+V(u)) + c_1^*a(t)(1+V(u)). \qquad [5.57]\end{aligned}$$

We denote by $\Theta_1, \Theta_2, \Theta_3$ the respective summands without coefficients in [5.56]. Consequently, for Θ_1, we have

$$\begin{aligned}\Theta_1 = \boldsymbol{C}_0(x)R_0\widetilde{\boldsymbol{L}}_t(x)V(u) &= C_0(u,x)R_0[\nabla_{b(t)}\widetilde{C}(u,x)V'(u)]'\\ &+ a(t)C_0(u,x)R_0[C_0(u,x)R_0[C_0(u,x)V'(u)]']'\\ &- a(t)\int_X \pi(dx)C_0(u,x)R_0[C_0(u,x)R_0[C_0(u,x)V'(u)]']'.\end{aligned}$$

Considering conditions C1–C8 of the theorem, we obtain

$$|\Theta_1| \le c_2^*(1 + V(u)). \qquad [5.58]$$

For Θ_2, we have

$$\Theta_2 = \nabla_{b(t)}\boldsymbol{C}(x)R_0\boldsymbol{C}_0(x)V(u) = \nabla_{b(t)}C(u,x)R_0[C_0(u,x)V'(u)]'.$$

In accordance with condition $C4$ of the theorem

$$|\Theta_2| \le c_2^*(1 + V(u)). \qquad [5.59]$$

For the third summand, we have

$$\begin{aligned}\Theta_3 &= \nabla_{b(t)}\boldsymbol{C}(x)R_0\widetilde{\boldsymbol{L}}_t(x)V(u)\\ &= \nabla_{b(t)}C(u,x)R_0[\nabla_{b(t)}\widetilde{C}(u,x)V'(u)]'\\ &\quad + a(t)\nabla_b C(u,x)R_0[C_0(u,x)R_0[C_0(u,x)V'(u)]']'\\ &\quad - a(t)\int\limits_X \pi(dx)\nabla_{b(t)}C(u,x)R_0[C_0(u,x)R_0[C_0(u,x)V'(u)]']'.\end{aligned}$$

Therefore,

$$|\Theta_3| \le c_3^*(1 + V(u)). \qquad [5.60]$$

Consequently, considering [5.58]–[5.60], for the remainder term $\Theta_t(x)V(u)$, we have the following common estimation:

$$|\Theta_t(x)V(u)| \le c_4^*(1 + V(u)). \qquad [5.61]$$

Considering [5.57] and [5.61], we obtain for $\boldsymbol{L}_t^\varepsilon V^\varepsilon(u,x)$ boundedness

$$\boldsymbol{L}_t^\varepsilon V^\varepsilon(u,x) \le -ca(t)V(u) + k * a(t)b(t)(1 + V(u)) + c *_5 a^2(t)(1 + V(u)).$$

From the last relation and the martingale characterizations of the process,

$$V^\varepsilon(u^\varepsilon(t), x_t^\varepsilon),\ t \ge O$$

follows the statement of the theorem by the Nevelson–Khasminsky theorem (Nevelson and Khasminsky 1972).

The theorem is thus proved.

COROLLARY 5.1.– *When perturbations $C_0(u,x) = C_0(x)$, i.e. it does not depend on the evolution u, the theorem holds true under the following conditions:*

*C3**: $|C_0(x)R_0[C_0(x)V''(u)]| \le c_2(1 + V(u)), c_2 > 0;$

*C4**: $|\nabla_{b(t)}C(u,x)R_0[C_0(x)V''(u)]| \le c_3(1 + V(u)), c_3 > 0;$

$C5*$:$|C_0(x)R_0[\nabla_{b(t)}\widetilde{C}(u,x)V'(u)]'| \leq c_4(1+V(u)), c_4 > 0;$

$C7*$:$|C_0(x)R_0[C_0(x)R_0[C_0(x)V'''(u)]]| \leq c_6(1+V(u)), c_6 > 0;$

$C8*$:$|\nabla_{b(t)}C(u,x)R_0[C_0(x)R_0[C_0(x)V'''(u)]]| \leq c_6(1+V(u)), c_6 > 0.$

COROLLARY 5.2.– *By calculating [5.57] the limit generator* $\mathbf{L}_t$, *we conclude the convergence of procedure [5.44] to the procedure that is defined by the stochastic differential equation*

$$du(t) = a(t)\nabla_{b(t)}C(u(t))dt + a^2(t)[b(u(t))dt + B(u(t))dw_t],$$

where the expression in square brackets is the perturbing diffusion process with a shift of function

$$b(t) = \Pi C_0(u,x)R_0C_0'(u,x) = \int_X \pi(dx)C_0(u,x)R_0C_0'(u,x),$$

and the variance matrix

$$B(u) = \Pi C_0(u,x)R_0C_0(u,x) = \int_X \pi(dx)C_0(u,x)R_0C_0(u,x);$$

$u \in R^d$ *and all the derivatives are calculated at the variable* u*:*

$$C_0'(u,x) = \partial C_0(u,x)/\partial u_k, k = \overline{1,N},$$

where w_t *is a standard Wiener process.*

5.2.2. *Fluctuations of the stochastic optimization procedure with diffusion perturbations*

The behavior of fluctuations for the stochastic optimization procedure characterizes the velocity of convergence of the system to the extremum point. We now consider the properties of fluctuations of the SOP with diffusion perturbations in the neighborhood of the extremum point of the averaged system. Furthermore, we investigate the asymptotic behavior of the SOP.

We consider the continuous procedure of stochastic optimization:

$$du^\varepsilon(t) = a(t)C^\varepsilon(u^\varepsilon(t), x(t/\varepsilon^4))dt, \quad [5.62]$$

with the regression function

$$C^\varepsilon(u,x) = \nabla_{b(t)}C(u,x) + \varepsilon^{-1}C_0(u,x), u \in R, x \in X \quad [5.63]$$

that satisfies the conditions of the existing global solution for accompanying systems:

$$\frac{u_x^\varepsilon(t)}{dt} = a(t)C^\varepsilon(u_x^\varepsilon(t), x), x \in X. \quad [5.64]$$

We also assume $C(u,\cdot) \in C^2(R)$, $C_0(u,\cdot) \in C^3(R)$.

The Markov process $x(t)$, $t \geq 0$ on a standard phase space $(X, \mathbf{X})$ is defined by the generator

$$\mathbf{Q}\varphi(x) = q(x) \int_X P(x, dy) \left[\varphi(y) - \varphi(x)\right], \quad \varphi \in \mathbf{B}(X),$$

where $\mathbf{B}(X)$ is a Banach space of bounded functions with the supremum norm $||\varphi|| = \max_{x \in X} |\varphi(x)|$.

Stochastic kernel $P(x, B), x \in X, B \in \mathbf{X}$ defines a uniformly ergodic embedded Markov chain $x_n = x(\tau_n)$, $n \geq 0$ with the stationary distribution $\rho(B), B \in \mathbf{X}$. The stationary distribution $\pi(B), B \in \mathbf{X}$ of the Markov process $x(t), t \geq 0$, is defined by the relation:

$$\pi(dx)q(x) = q\rho(dx), \; q = \int_X \pi(dx)q(x).$$

The potential operator R_0 of the generator $\mathbf{Q}$ is defined by the relation:

$$R_0 = \Pi - (\Pi + \mathbf{Q})^{-1},$$

where $\Pi\varphi(x) = \int_X \pi(dy)\varphi(y)$ is a projector onto the subspace $N_Q = \{\varphi : \mathbf{Q}\varphi = 0\}$ of zeros of the operator $\mathbf{Q}$.

For the perturbation $C_0(u, x)$ of the regression function, let the balance condition hold true

$$\int_X \pi(dx)C_0(u, x) \equiv 0. \qquad [5.65]$$

Under the conditions of Theorem 5.5, continuous SOP [5.44] with probability 1 converges to the extremum of the averaged system

$$\frac{du(t)}{dt} = C'(u(t)), \qquad [5.66]$$

where

$$C(u) = \int_X \pi(dx)C(u, x).$$

The condition of the existing extremum point u_0 is the balance condition

$$\Pi C'(0, x) = \int_X \pi(dx)C'(0, x) = 0. \qquad [5.67]$$

The control function $a(t)$ is defined by the relation

$$a(t) = a/t, 0 < t_0 < t, a > 0.$$

The diffusion perturbation is defined by the relation

$$\bar{C}_0^\varepsilon(t) = \varepsilon^{-2} a \int_{t_0}^{t} C_0(u^\varepsilon(s), x(s/\varepsilon^4))/s ds.$$

We denote $C_0^0(x) = C_0(0, x)$, then the diffusion perturbation at the extremum point has the form

$$C_0^\varepsilon(t) = \varepsilon^{-2} a \int_{t_0}^{t} C_0(0, x(s/\varepsilon^4))/s ds = \varepsilon^{-2} a \int_{t_0}^{t} C_0^0(x(s/\varepsilon^4))/s ds. \quad [5.68]$$

We investigate the fluctuation of the SOP in the form

$$v^\varepsilon(t) = \varepsilon^{-1} t^{1/2} \left[u^\varepsilon(t) - \varepsilon C_0^\varepsilon(t)\right]. \quad [5.69]$$

THEOREM 5.6.– *Let the balance conditions [5.67] and [5.45], the convergence conditions of SOP [5.62] hold true, and the functions $C_0'(u, x)$ and $C_0''(u, x)$ be uniformly bounded at x*

$$\sup_{x \in X} ||C_0'(u, x)|| \leq C_1 < +\infty;$$

$$\sup_{x \in X} ||C_0''(u, x)|| \leq C_2 < +\infty. \quad [5.70]$$

Then, the weak convergence

$$(v^\varepsilon(t), C_0^\varepsilon(t)) \to (\zeta(t), W_\sigma(t)), t > 0, \qquad \varepsilon \to 0$$

holds true at each finite interval $0 < t_0 < t < T$.

The limit process $(\zeta(t), W_\sigma(t))$ *is given by the generator*

$$\boldsymbol{L}\varphi(v, w) = \left[\frac{acv^2}{\sqrt{t}} + acvw + \frac{v}{2} + \frac{ac\sqrt{t}w^2}{2}\right] \varphi_v'(v, w) + \frac{a^2\rho^2}{2t} \varphi_w''(v, w),$$

where

$$c = \int_X \pi(dx) C''(0, x);$$

$$\rho^2 = 2 \int_X \pi(dx) C_0(0, x) R_0 C_0(0, x).$$

PROOF OF THE THEOREM.– We will consider additional lemmas.

LEMMA 5.15.– *The fluctuation [5.69] satisfies the stochastic differential equation*

$$dv^{\varepsilon}(t) = C_t^{\varepsilon}(v^{\varepsilon}(t), x_t^{\varepsilon})dt, \qquad [5.71]$$

where

$$C_t^{\varepsilon}(v,x) = \frac{\varepsilon^{-2}a}{t^{1/2}}\left[C_0(\varepsilon z^{\varepsilon}(t),x) - C_0(0,x)\right] + \frac{\varepsilon^{-1}a}{t^{1/2}}\nabla_{b(t)}C(\varepsilon z^{\varepsilon}(t),x) + \frac{v}{2t},$$

$$z^{\varepsilon}(t) = t^{-1/2}v^{\varepsilon}(t) + w,$$
$$w = C_0^{\varepsilon}(t),$$
$$v = v^{\varepsilon}(t).$$

PROOF.–

Differentiating [5.68], we obtain

$$dv^{\varepsilon}(t) = \varepsilon^{-1}\frac{1}{2}t^{-\frac{1}{2}}\left[u^{\varepsilon}(t) - \varepsilon C_0^{\varepsilon}(t)\right]dt + \varepsilon^{-1}t^{1/2}\left[du^{\varepsilon}(t) - \varepsilon dC_0^{\varepsilon}(t)\right]. \quad [5.72]$$

Since

$$du^{\varepsilon}(t) = \frac{a}{t}\left[\nabla_{b(t)}C(u^{\varepsilon}(t),x) + \varepsilon^{-1}C_0(u^{\varepsilon}(t),x)\right]dt,$$

and

$$dC_0^{\varepsilon}(t) = \varepsilon^{-2}at^{-1}C_0(0,x)dt,$$

for [5.72], we then have

$$\begin{aligned} dv^{\varepsilon}(t) = {} & \frac{1}{2t}v^{\varepsilon}(t)dt + \varepsilon^{-1}t^{1/2}[\frac{a}{t}[\nabla_{b(t)}C(\varepsilon z^{\varepsilon}(t),x) + \varepsilon^{-1}C_0(\varepsilon z^{\varepsilon}(t),x)] \\ & - \varepsilon^{-1}\frac{a}{t}C_0^{\varepsilon}(t)dt] = [\varepsilon^{-2}at^{-1/2}C_0(\varepsilon z^{\varepsilon}(t),x) \\ & + \varepsilon^{-1}at^{-1/2}\nabla_{b(t)}C(\varepsilon z^{\varepsilon}(t),x) - \varepsilon^{-2}at^{-1/2}C_0(0,x) + \frac{1}{2t}v^{\varepsilon}(t)]dt. \end{aligned}$$

From the last relation, we obtain the expected statement.

For the system [5.71], we consider the parametrical system

$$dv_x^{\varepsilon}(t) = C_t^{\varepsilon}(v_x^{\varepsilon}(t), x)dt \qquad [5.73]$$

with the semigroup of the operators

$$\mathbf{C}_t^{\varepsilon}(x)\varphi(v) = \varphi(v_x^{\varepsilon}(t+s)), v_x^{\varepsilon}(s) = v,$$

on the solutions of the system [5.73] with the generator

$$\mathbf{C}_t^{\varepsilon,V}(x)\varphi(v) = C_t^{\varepsilon}(v,x)\varphi'(v). \qquad [5.74]$$

LEMMA 5.16.– *The generator [5.74] has the asymptotic representation*

$$\mathbf{C}_t^{\varepsilon,V}(x)\varphi(v) = \varepsilon^{-1}\frac{a}{t^{1/2}}[z^\varepsilon(t)C_0'(0,x) + C'(0,x)]\varphi'(v)$$
$$+ \frac{a}{t^{1/2}}z^\varepsilon(t)[\frac{z^\varepsilon(t)}{2}C_0''(0,x) + C''(0,x)]\varphi'(v)$$
$$+ \frac{1}{2t}v\varphi'(v) + \theta^\varepsilon(x)\varphi(v),$$

where $\theta^\varepsilon(x)\varphi(v)$ is the remainder term and

$$|\theta^\varepsilon(x)\varphi(v)| \to 0$$

at $\varepsilon \to 0$.

PROOF.– From [5.74], we have

$$\mathbf{C}_t^{\varepsilon,V}(x)\varphi(v) = C^\varepsilon(v,x)\varphi'(v)$$
$$= [\varepsilon^{-2}\frac{a}{t^{1/2}}[C_0(\varepsilon z^\varepsilon(t),x) - C_0(0,x)]$$
$$+ \varepsilon^{-1}\frac{a}{t^{1/2}}\nabla_{b(t)}C(\varepsilon z^\varepsilon(t),x) + \frac{1}{2t}v]\varphi'(v)$$
$$= [\varepsilon^{-2}\frac{a}{t^{1/2}}[C_0(0,x) + \varepsilon z^\varepsilon(t)C_0'(0,x) + \frac{\varepsilon^2 z^\varepsilon(t)^2}{2}C_0''(0,x)$$
$$+ O(\varepsilon^3) - C_0(0,x)]\varphi'(v)$$
$$+ \varepsilon^{-1}\frac{a}{t^{1/2}}[C'(0,x) + \varepsilon z^\varepsilon(t)C''(0,x) + O(\varepsilon^3)] + \frac{1}{2t}v]\varphi'(v)$$
$$= [\varepsilon^{-1}\frac{a}{t^{1/2}}[z^\varepsilon(t)C_0'(0,x) + C'(0,x)]$$
$$+ \frac{a}{t^{1/2}}z^\varepsilon(t)[\frac{z^\varepsilon(t)}{2}C_0''(0,x) + C''(0,x)]$$
$$+ \frac{1}{2t}v]\varphi'(v) + \varepsilon\theta_v^\varepsilon(x)\varphi'(v),$$

where $\theta_v^\varepsilon(x)\varphi'(v)$ is the uniformly bounded function at x.

LEMMA 5.17.– *The generator $\mathbf{L}^\varepsilon$ of the three-component Markov process*

$$v^\varepsilon(t) =: v_t,\ C_0^\varepsilon(t) =: w_t,\ x(t/\varepsilon^4) = x_t^\varepsilon,\ t \geq 0 \qquad [5.75]$$

on the test functions $\varphi(v, w, \cdot) \in C^{2,2}(R^d \times R^d)$ *has the asymptotic representation*

$$\begin{aligned} \boldsymbol{L}^{\varepsilon}\varphi(v,w,x) &= \varepsilon^{-4}\mathbf{Q}\varphi(v,w,x) + \varepsilon^{-2}\frac{a}{t}\boldsymbol{C}_0(x)\varphi(v,w,x) \\ &+ \varepsilon^{-1}\frac{a}{t^{1/2}}\boldsymbol{C}_0^1(v,w,x)\varphi(v,w,x) + \frac{a}{t}\boldsymbol{C}_0^2(v,w,x)\varphi(v,w,x) \\ &+ \theta^{\varepsilon}_{L^{\varepsilon}}(x)\varphi(v,w,x), \end{aligned} \quad [5.76]$$

where

$$\begin{aligned} &\mathbf{C}_0(x)\varphi(v,w,x) = C_0(0,x)\varphi'_w(v,w,x). \\ &\boldsymbol{C}_0^1(v,w,x)\varphi(v,w,x) = (zC_0'(0,x) + C'(0,x))\varphi'(v,w,x), \\ &\boldsymbol{C}_0^2(v,w,x)\varphi(v,w,x) = (\sqrt{t}z(\frac{z}{2}C_0''(0,x) + C''(0,x)) + \frac{v}{2a})\varphi'(v,w,x), \end{aligned}$$

and the remainder term $||\theta^{\varepsilon}_{L^{\varepsilon}}(x)\varphi(v,w,x)|| \to 0$ *at* $\varepsilon \to 0$.

PROOF.– In accordance with the definition of the generator of the Markov process, we have

$$\begin{aligned} \boldsymbol{L}^{\varepsilon}\varphi(v,w,x) &= \lim_{\Delta\to 0}\frac{1}{\Delta}E[\varphi(v^{\varepsilon}(t+\Delta), C_0^{\varepsilon}(t+\Delta), x((t+\Delta)/\varepsilon^4)) \\ &\quad - \varphi(v^{\varepsilon}(t), C_0^{\varepsilon}(t), x(t/\varepsilon^4))|v^{\varepsilon}(t)=v, C_0^{\varepsilon}(t)=w, x(t/\varepsilon^4)=x] \\ &= \lim_{\Delta\to 0}\frac{1}{\Delta}E_{v,w,x}[\varphi(v_{t+\Delta}, w_{t+\Delta}, x_{t+\Delta}) - \varphi(v,w,x)] \\ &= \lim_{\Delta\to 0}\frac{1}{\Delta}E_{v,w,x}[\varphi(v,w,x_{t+\Delta}) - \varphi(v,w,x)] \\ &\quad + \lim_{\Delta\to 0}\frac{1}{\Delta}E_{v,w,x}[\varphi(v_{t+\Delta}, w_{t+\Delta}, x_{t+\Delta}) - \varphi(v_{t+\Delta}, w, x_{t+\Delta})] \\ &\quad + \lim_{\Delta\to 0}\frac{1}{\Delta}E_{v,w,x}[\varphi(v_{t+\Delta}, w, x_{t+\Delta}) - \varphi(v,w,x_{t+\Delta})]. \end{aligned}$$

By the relations

$$\begin{aligned} &\lim_{\Delta\to 0}\frac{1}{\Delta}E_{v,w,x}[\varphi(v,w,x_{t+\Delta}) - \varphi(v,w,x)] = \varepsilon^{-4}\mathbf{Q}\varphi(v,w,x), \\ &\lim_{\Delta\to 0}\frac{1}{\Delta}E_{v,w,x}[\varphi(v_{t+\Delta}, w_{t+\Delta}, x_{t+\Delta}) - \varphi(v_{t+\Delta}, w, x_{t+\Delta})] \\ &= \lim_{\Delta\to 0}\frac{1}{\Delta}E_{v,w,x}[\varphi'_w(v_{t+\Delta}, w, x_{t+\Delta})]\frac{dC_0^{\varepsilon}(t)}{dt} \\ &= \varepsilon^{-2}\frac{a}{t}C_0(0,x)\varphi'_w(v,w,x), \\ &\lim_{\Delta\to 0}\frac{1}{\Delta}E_{v,w,x}[\varphi(v_{t+\Delta}, w, x_{t+\Delta}) - \varphi(v,w,x_{t+\Delta})] = \mathbf{C}_t^{\varepsilon,V}(x)\varphi(v,w,x), \end{aligned}$$

considering Lemma 5.16, we obtain

$$\begin{aligned}\boldsymbol{L}^{\varepsilon}\varphi(v,w,x) &= \varepsilon^{-4}\mathbf{Q}\varphi(v,w,x)\\ &+\varepsilon^{-2}\frac{a}{t}C_0(0,x)\varphi'_w(v,w,x)\\ &+\varepsilon^{-1}\frac{a}{t^{1/2}}[zC'_0(0,x)+C'(0,x)]\varphi'_v(v,w,x)\\ &+\frac{a}{t^{1/2}}z[\frac{z}{2}C''_0(0,x)+C''(0,x)]\varphi'_v(v,w,x)\\ &+\frac{1}{2t}v\varphi'_v(v,w,x)+\theta^{\varepsilon}_{L^{\varepsilon}}(x)\varphi(v,w,x)\end{aligned}$$

and the lemma is proved.

Furthermore, we consider the truncated operator $\boldsymbol{L}^{\varepsilon}$

$$\boldsymbol{L}_0^{\varepsilon} = \varepsilon^{-4}\mathbf{Q}+\varepsilon^{-2}\frac{a}{t}\boldsymbol{C}_0(x)+\varepsilon^{-1}\frac{a}{t^{1/2}}\boldsymbol{C}_0^1(v,w,x)+$$
$$+\frac{a}{t}\boldsymbol{C}_0^2(v,w,x). \qquad [5.77]$$

To construct the limit operator, we find the solution of the singular perturbation problem for the operator [5.77] on the test function in the form

$$\begin{aligned}\varphi^{\varepsilon}(v,w,x) &= \varphi(v,w)+\varepsilon^2\frac{1}{t}\varphi_2(v,w,x)\\ &+\varepsilon^3\frac{1}{\sqrt{t}}\varphi_3(v,w,x)+\varepsilon^4\frac{1}{t}\varphi_4(v,w,x).\end{aligned} \qquad [5.78]$$

LEMMA 5.18.– *The solution of the singular perturbation problem for the operator [5.77] on the test function [5.78] has the form*

$$\boldsymbol{L}_0^{\varepsilon}\varphi^{\varepsilon}(v,w,x) = \frac{1}{t}\boldsymbol{L}\varphi(v,w)+\varepsilon\theta^{\varepsilon}_{L_0^{\varepsilon}}(v,w,x)\varphi(v,w), \qquad [5.79]$$

where the remainder term $\theta^{\varepsilon}_{L_0^{\varepsilon}}(v,w,x)\varphi(v,w)$ *has the representation*

$$\begin{aligned}\theta^{\varepsilon}_{L_0^{\varepsilon}}(v,w,x)\varphi(v,w) &= \frac{a}{t^{3/2}}\theta_1^{\varepsilon}(v,w,x)+\frac{a}{t}\theta_2^{\varepsilon}(v,w,x)+\frac{a}{t^2}\theta_3^{\varepsilon}(v,w,x),\\ \theta_1^{\varepsilon}(v,w,x) &= \boldsymbol{C}_0(x)\varphi_3(v,w,x)\\ &+\boldsymbol{C}_0^1(v,w,x)\varphi_2(v,w,x)+\varepsilon^2\boldsymbol{C}_0^1(v,w,x)\varphi_4(v,w,x)\\ &+\varepsilon^2\boldsymbol{C}_0^2(v,w,x)\varphi_3(v,w,x),\\ \theta_2^{\varepsilon}(v,w,x) &= \varepsilon\boldsymbol{C}_0^1(v,w,x)\varphi_3(v,w,x),\\ \theta_3^{\varepsilon}(v,w,x) &= \varepsilon\boldsymbol{C}_0(x)\varphi_4(v,w,x)+\varepsilon\boldsymbol{C}_0^2(v,w,x)\varphi_2(v,w,x)\\ &+\varepsilon^3\boldsymbol{C}_0^2(v,w,x)\varphi_4(v,w,x),\end{aligned}$$

and the limit operator $\mathbf{L}$ *has the form:*

$$\mathbf{L}\Pi = a\Pi\mathbf{C}_0^2(v,w,x)\Pi + \frac{a^2}{t}\Pi\mathbf{C}_0(x)R_0\mathbf{C}_0(x)\Pi. \quad [5.80]$$

PROOF.– We consider the operator $\mathbf{L}_0^\varepsilon$ on the test function [5.80]:

$$\begin{aligned}\mathbf{L}_0^\varepsilon\varphi^\varepsilon(v,w,x) &= [\varepsilon^{-4}\mathbf{Q} + \varepsilon^{-2}\frac{a}{t}\boldsymbol{C}_0(x) + \varepsilon^{-1}\frac{a}{t^{1/2}}\boldsymbol{C}_0^1(v,w,x)\\ &\quad + \frac{a}{t}\boldsymbol{C}_0^2(v,w,x)][\varphi(v,w) + \varepsilon^2\frac{1}{t}\varphi_2(v,w,x)\\ &\quad + \varepsilon^3\frac{1}{\sqrt{t}}\varphi_3(v,w,x) + \varepsilon^4\frac{1}{t}\varphi_4(v,w,x)]\\ &= \varepsilon^{-4}\mathbf{Q}\varphi(v,w) + \varepsilon^{-2}\frac{1}{t}[\mathbf{Q}\varphi_2(v,w,x) + a\boldsymbol{C}_0(x)\varphi(v,w)]\\ &\quad + \varepsilon^{-1}\frac{1}{t^{1/2}}[\mathbf{Q}\varphi_3(v,w,x) + a\boldsymbol{C}_0^1(v,w,x)\varphi(v,w)]\\ &\quad + \frac{1}{t}[a\boldsymbol{C}_0^2(v,w,x)\varphi(v,w) + \frac{a}{t}\mathbf{C}_0(x)\varphi_2(v,w,x)\\ &\quad + \mathbf{Q}\varphi_3(v,w,x)] + \varepsilon\boldsymbol{\theta}_{L_0^\varepsilon}^\varepsilon(v,w,x)\varphi(v,w).\end{aligned}$$

The first condition of solvability holds true, because $\varphi(v,w)$ does not depend on x, i.e. we have

$$\mathbf{Q}\varphi(v,w) = 0.$$

The second condition of solvability of the singular perturbation problem

$$\mathbf{Q}\varphi_2(v,w,x) + a\boldsymbol{C}_0(x)\varphi(v,w) = 0$$

holds true, because the balance condition [5.45] is fulfilled. From the last relation, we then have

$$\varphi_2(v,w,x) = aR_0\boldsymbol{C}_0(x)\varphi(v,w). \quad [5.81]$$

The second condition of solvability gives the equation

$$\mathbf{Q}\varphi_3(v,w,x) + a\boldsymbol{C}_0^1(v,w,x)\varphi(v,w) = 0,$$

and along with the balance condition, we have

$$\varphi_3(v,w,x) = aR_0\boldsymbol{C}_0^1(v,w,x)\varphi(v,w). \quad [5.82]$$

The fourth condition has the form

$$\frac{1}{t}[\mathbf{Q}\varphi_4(v,w,x) + \boldsymbol{L}(x)\varphi(v,w)] = \frac{1}{t}\boldsymbol{L}\varphi(v,w),$$

where

$$\boldsymbol{L}(x)\varphi(v,w) = a\boldsymbol{C}_0^2(v,w,x)\varphi(v,w) + \frac{a}{t}\boldsymbol{C}_0(x)\varphi_2(v,w,x)$$
$$= a\boldsymbol{C}_0^2(v,w,x)\varphi(v,w) + \frac{a^2}{t}\boldsymbol{C}_0(x)R_0\boldsymbol{C}_0(x)\varphi(v,w). \qquad [5.83]$$

The limit operator has the form

$$\boldsymbol{L} = \Pi\boldsymbol{L}(x)\Pi,$$

and therefore, in accordance with [5.83], we obtain [5.80].

Consequently, for the component $\varphi_4(v,w,x)$, we have the representation

$$\varphi_4(v,w,x) = R_0\widetilde{\boldsymbol{L}}(x)\varphi(v,w), \qquad [5.84]$$

where

$$\widetilde{\boldsymbol{L}}(x) = \boldsymbol{L}(x) - \boldsymbol{L}.$$

Finally, calculating the relation [5.84]

$$\mathbf{L}\varphi(v,w) = a[\sqrt{t}z(\frac{z}{2}\int_X \pi(dx)(C_0''(0,x) + C''(0,x))) + \frac{v}{2a}]\varphi_v'(v,w)$$
$$+\frac{a^2}{t}\int_X \pi(dx)C_0(0,x)R_0C_0(0,x)\varphi_w''(v,w).$$

by the balance condition [5.45], we have

$$\int_X \pi(dx)C_0''(0,x) = 0,$$

therefore, we finally have

$$\mathbf{L}\varphi(v,w) = (a\sqrt{t}\frac{z^2}{2}c + \frac{v}{2})\varphi_v'(v,w) +$$
$$+\frac{a^2\rho^2}{2t}\varphi_w''(v,w),$$

under the notation of the theorem. Since

$$z = t^{-1/2}v + w,$$

from the last relation, we have the representation of the limit generator.

It remains to prove the boundedness of the remainder term

$$||\theta^{\varepsilon}_{L^{\varepsilon}_0}(v,w,x)\varphi(v,w)|| \leq M, 0 < M < \infty \qquad [5.85]$$

in relation [5.79]. From the relations [5.80], [5.81] and [5.83] for $\theta^{\varepsilon}_1(v,w,x)$, we have

$$\begin{aligned}\theta^{\varepsilon}_1(v,w,x) &= a\boldsymbol{C}_0(x)R_0\boldsymbol{C}^1_0(v,w,x)\varphi(v,w)\\ &+ a\boldsymbol{C}^1_0(v,w,x)R_0\boldsymbol{C}_0(x)\varphi(v,w)\\ &+ \varepsilon^2\boldsymbol{C}^1_0(v,w,x)R_0\widetilde{\boldsymbol{L}}(x)\varphi(v,w)\\ &+ \varepsilon a^2\boldsymbol{C}^2_0(v,w,x)R_0\boldsymbol{C}^1_0(v,w,x)\varphi(v,w).\end{aligned} \qquad [5.86]$$

For the first term from [5.86], we have

$$\boldsymbol{C}_0(x)R_0\boldsymbol{C}^1_0(v,w,x)\varphi(v,w)$$

$$= C_0(0,x)R_0[((\frac{v}{\sqrt{t}}w)C'_0(0,x) + C'(0,x))\varphi'_v(v,w)]'_w.$$

The second term has the representation

$$\boldsymbol{C}^1_0(v,w,x)R_0\boldsymbol{C}_0(x)\varphi(v,w)$$

$$= ((\frac{v}{\sqrt{t}} + w)C'_0(0,x) + C'(0,x))R_0[C_0(0,x)\varphi'_w(v,w)]'_v.$$

For the third term, we have

$$\boldsymbol{C}^1_0(v,w,x)R_0\widetilde{\boldsymbol{L}}(x)\varphi(v,w)$$

$$= ((v+\sqrt{t}w)(\frac{v+\sqrt{t}w}{2\sqrt{t}}C''_0(0,x) + C''(0,x)) + \frac{v}{2a}R_0[\widetilde{\boldsymbol{L}}(x)\varphi(v,w)]'_v.$$

The fourth summand has the representation

$$\boldsymbol{C}^2_0(v,w,x)R_0\boldsymbol{C}^1_0(v,w,x)\varphi(v,w)$$

$$= ((v+\sqrt{t}w)(\frac{v+\sqrt{t}w}{2\sqrt{t}}C''_0(0,x) + C''(0,x)) + \frac{v}{2a})$$

$$R_0[(\frac{v+\sqrt{t}w}{2\sqrt{t}}C'_0(0,x) + C'(0,x))\varphi'_v(v,w)]'_v.$$

By the smoothness of the functions $C_0(u,x)$, $C(u,x)$ and the test functions $\varphi(v,w)$ from the obtained representations of terms and from [5.86], we have the estimation

$$||\theta^{\varepsilon}_1(v,w,x)|| \leq M_1, 0 < M_1 < \infty.$$

There are also estimations

$$||\theta_2^\varepsilon(v, w, x)|| \leq M_2, 0 < M_2 < \infty.$$

$$||\theta_3^\varepsilon(v, w, x)|| \leq M_3, 0 < M_3 < \infty.$$

Therefore, [5.85] holds true.

According to the Korolyuk model theorem, we obtain the statement of Theorem 5.6.

5.2.3. *Fluctuation of the SOP*

The difference stochastic optimization procedure in an ergodic Markov environment with impulse perturbation is defined by the evolutionary equation

$$du^\varepsilon(t) = a(t)[\nabla_{b(t)} C(u^\varepsilon(t), x(t/\varepsilon^2))dt + \varepsilon d\eta^\varepsilon(t)], \qquad [5.87]$$

where

$$a(t) = \frac{a}{t}, a > 0,$$

$$\nabla_{b(t)} C(u, x) = \frac{1}{2b(t)}[C(u + b(t), x) - C(u + b(t), x)], \; C(u, \cdot) \in C^3(R)).$$

The Markov process $x(t), t \geq 0$ on a standard phase space $(X, \mathbf{X})$ is given by the generator:

$$\mathbf{Q}\varphi(x) = q(x) \int_X P(x, dy) [\varphi(y) - \varphi(x)], \quad \varphi \in \mathbf{B}(X),$$

where $\mathbf{B}(X)$ is the Banach space of real-valued bounded functions with the supremum norm $||\varphi|| = \max_{x \in X} |\varphi(x)|$.

The stochastic kernel $P(x, B), x \in X, B \in \mathbf{X}$, defines a uniformly ergodic embedded Markov chain $x_n = x(\tau_n), n \geq 0$ with the stationary distribution $\rho(B), B \in \mathbf{X}$.

The stationary distribution $\pi(B), B \in \mathbf{X}$, of the Markov process $x(t), t \geq 0$, is defined by the relation

$$\pi(dx)q(x) = q\rho(dx), \; q = \int_X \pi(dx)q(x).$$

The potential operator R_0 of the generator $\mathbf{Q}$ is defined by the relation: $R_0 = \Pi - (\Pi + \mathbf{Q})^{-1}$, where $\Pi\varphi(x) = \int_X \pi(dy)\varphi(y)$ is a projector onto the subspace $N_Q = \{\varphi : \mathbf{Q}\varphi = 0\}$ of zeros of the operator $\mathbf{Q}$.

The impulse perturbation process (IPP) $\eta^\varepsilon(t), t \geq 0$, is given by the relation:

$$\eta^\varepsilon(t) = \int_0^t \eta^\varepsilon(ds; x(s/\varepsilon^2)); \qquad [5.88]$$

where the family of processes with independent increments $\eta^\varepsilon(t, x), t \geq 0, x \in X$, is given by the generator

$$\mathbf{\Gamma}_w^\varepsilon(x)\varphi(w) = \varepsilon^{-2} \int_R [\varphi(w + \varepsilon v) - \varphi(w)]\Gamma(dv; x), x \in X.$$

We assume that the balance condition is satisfied:

$$\Pi\mathbf{\Gamma}_1(x) = \int_X \pi(dx)b_1(x) = 0, b_1(x) = \int_R v\Gamma(dv, x), \qquad [5.89]$$

$$\Pi C'(0, x) = \int \pi(dx)C'(0, x) = 0. \qquad [5.90]$$

The normalized impulse perturbation is given by the relation

$$\mu^\varepsilon(t) = a \int_{t_0}^t \eta^\varepsilon(ds; x(s/\varepsilon^4))/s^{\frac{1}{2}}.$$

We consider the fluctuation of the SOP in the form

$$v^\varepsilon(t) = \varepsilon^{-1}t^{\frac{1}{2}}[u^\varepsilon(t) - \varepsilon\mu^\varepsilon(t)]. \qquad [5.91]$$

THEOREM 5.7.– *Under the balance conditions [5.89] and [5.90], as well as the convergence condition for the SOP, the weak convergence*

$$(v^\varepsilon(t), \mu^\varepsilon(t)) \to (\zeta(t), \sigma(t)W(t)), t > 0, \qquad \varepsilon \to 0$$

holds true at every finite interval $0 < t_0 < t < T$.

The limit process $(\zeta(t), \sigma(t)W(t))$ *is defined by the generator*

$$\mathbf{L}\varphi(v, w) = \left[v(ac_2 + \frac{1}{2}t^{-\frac{1}{2}}) + act^{\frac{1}{2}}w\right] \varphi_v'(v, w) + \frac{a}{2}B(t)\varphi_w''(v, w),$$

where

$$c_2 = \int_X \Pi(dx) C''(0,x),$$

$$B(t) = B_1(t) + B_2(t) + B_3(t),$$

$$B_1(t) = \int_X \pi(dx) b_2(x) + 2at^{-\frac{1}{2}} \int_X \pi(dx) b_1(x) R_0 b_1(x);$$

$$b_2(x) = \int_R v^2 \Gamma(dv, x);$$

$$B_2(t) = 2a \left[\int_X \pi(dx) b_1(x) R_0 C'(0,x) + \int_X \pi(dx) C'(0,x) R_0 b_1(x) \right];$$

$$B_3(t) = 2at^{\frac{1}{2}} \int_X \pi(dx) C'(0,x) R_0 C'(0,x).$$

LEMMA 5.19.– *The generator* $\mathbf{\Gamma}^\varepsilon(x)\varphi(w)$ *has the asymptotic representation*

$$\mathbf{\Gamma}^\varepsilon(x)\varphi(w) = \varepsilon^{-2}\mathbf{\Gamma_1}(x)\varphi(w) + \mathbf{\Gamma_2}(x)\varphi(w) + \varepsilon\gamma^\varepsilon(x)\varphi(w),$$

where $\mathbf{\Gamma_1}(x)\varphi(w) = b_1(x)\varphi'(w)$; $\mathbf{\Gamma_2}(x)\varphi(w) = \frac{1}{2}b_2(x)\varphi''(w)$, *and the remainder term* $||\varepsilon\gamma^\varepsilon(x)\varphi(w)|| \to 0$ *at* $\varepsilon \to 0$.

PROOF.– By the definition of the generator, we have

$$\mathbf{\Gamma}^\varepsilon(x)\varphi(w) = \varepsilon^{-2} \int_R [\varphi(w + \varepsilon v) - \varphi(w)] \Gamma(dv, x).$$

Considering the decomposition of the function φ into the Taylor series, we obtain

$$\begin{aligned}
\mathbf{\Gamma}^\varepsilon(x)\varphi(w) &= \varepsilon^{-2} \int_R [\varphi(w) + \varepsilon v \varphi'(w) + \frac{1}{2}\varepsilon^2 v^2 \varphi''(w) \\
&\quad + \varepsilon\gamma^\varepsilon(x)\varphi(w) - \varphi(w)]\Gamma(dv,x) = \varepsilon^{-2} \int_R (\varepsilon v \varphi'(w)\Gamma(dv,x) \\
&\quad + \frac{1}{2}\varepsilon^{-2} \int_R \varepsilon^2 v^2 \varphi''(w)\Gamma(dv,x) + \varepsilon\gamma^\varepsilon(x)\varphi(w) \\
&= \varepsilon^{-1} b_1(x)\varphi'(w) + \frac{1}{2} b_2(x)\varphi''(w) + \varepsilon\gamma^\varepsilon(x)\varphi(w) \\
&= \varepsilon^{-1}\mathbf{\Gamma}_1(x)\varphi(w) + \mathbf{\Gamma}_2(x)\varphi(w) + \varepsilon\gamma^\varepsilon(x)\varphi(w).
\end{aligned}$$

LEMMA 5.20.– *The normalized fluctuation [5.91] satisfies the differential equation*

$$dv^\varepsilon(t) = C_t^{\nabla,\varepsilon}(v^\varepsilon(t), x_t^\varepsilon)dt, \qquad [5.92]$$

where

$$C_t^{\nabla,\varepsilon}(v,x) = \varepsilon^{-1}a\nabla_{b(t)}C(\varepsilon z, x) + \frac{1}{2}vt^{-1},$$

and

$$z = t^{-\frac{1}{2}}v + w; v = v^\varepsilon(t), w = \mu^\varepsilon(t).$$

PROOF.–

Differentiating [5.91] and using [5.86], we obtain [5.92].

We note that equation [5.92] generates the semigroups of the operators $\mathbf{C}_t^\varepsilon(x)\varphi(v) = \varphi(v_x^\varepsilon(t+s))$, $v_x^\varepsilon(s) = v$, with the generator

$$\mathbf{C}_t^{\nabla,\varepsilon}(x)\varphi(v) = C^{\nabla,\varepsilon}(v,x)\varphi'(v). \qquad [5.93]$$

LEMMA 5.21.– *The generator [5.93] has the asymptotic representation of the test functions*

$$\varphi(v) \in C^2(R), C(u,\cdot) \in C^3(R)$$

$$\mathbf{C}_t^{\nabla,\varepsilon}(x)\varphi(v) = \varepsilon^{-1}aC'(0,x)\varphi'(v) + ((avt^{-\frac{1}{2}} + aw)C''(0,x)$$
$$+ \frac{1}{2}t^{-1}v)\varphi'(v) + \varepsilon\theta^\varepsilon(x)\varphi'(v), \qquad [5.94]$$

where $\theta^\varepsilon(x)\varphi'(v)$ is the function uniformly bounded at x and v.

PROOF.– For the generator [5.93], there are some transformations:

$$\mathbf{C}_t^{\nabla,\varepsilon}(x)\varphi(v) = \varepsilon^{-1}a\nabla_{b(t)}C(\varepsilon z, x)\varphi'(v) + \frac{1}{2}vt^{-1}\varphi'(v)$$
$$= \varepsilon^{-1}a\left[C'(0,x) + \varepsilon zC''(0,x) + O(\varepsilon^3)\right]\varphi'(v) + \frac{1}{2}vt^{-1}\varphi'(v).$$

Therefore, reducing the terms by degrees of ε, we obtain [5.94].

LEMMA 5.22.– *The generator of the three-component Markov process*

$$v^\varepsilon(t) = v_t,\ \mu^\varepsilon(t) = w_t,\ x(t/\varepsilon^2) = x_t^\varepsilon,\ t \geq 0 \qquad [5.95]$$

has the analytical form

$$\mathbf{L}^\varepsilon\varphi(v,w,x) = \left[\varepsilon^{-2}\mathbf{Q} + at^{-\frac{1}{2}}\mathbf{\Gamma}^\varepsilon(x) + \mathbf{C}_t^{\nabla,\varepsilon}(x)\right]\varphi(v,w,x).$$

PROOF.– We calculate the conditional expectation:

$$\begin{aligned}
&E[\varphi(v^\varepsilon(t+\Delta), \mu^\varepsilon(t+\Delta), x((t+\Delta)/\varepsilon^2)) \\
&\quad -\varphi(v^\varepsilon(t), \mu^\varepsilon(t), x(t/\varepsilon^2))|v^\varepsilon(t) = v, \mu^\varepsilon(t) = w, x(t/\varepsilon^2) = x] \\
&= E[\varphi(v_{t+\Delta}, w_{t+\Delta}, x_{t+\Delta}) - \varphi(v, w, x)] \\
&= E[\varphi(v_{t+\Delta}, w_{t+\Delta}, x_{t+\Delta}) - \varphi(v, w_{t+\Delta}, x_{t+\Delta})] \\
&\quad +E[\varphi(v, w_{t+\Delta}, x_{t+\Delta}) - \varphi(v, w, x)].
\end{aligned}$$

In accordance with the definition of the generator of the Markov process and the last representation, we obtain

$$\begin{aligned}
\mathbf{L}^\varepsilon\varphi(v, w, x) &= \lim_{\Delta\to 0}\frac{1}{\Delta}E[\varphi(v^\varepsilon(t+\Delta), \mu^\varepsilon(t+\Delta), x((t+\Delta)/\varepsilon^2)) \\
&\quad - v^\varepsilon(t), \mu^\varepsilon(t), x(t/\varepsilon^2))|v^\varepsilon(t) = v, \mu^\varepsilon(t) = w, x(t/\varepsilon^2) = x] \\
&= \lim_{\Delta\to 0}\frac{1}{\Delta}E[v_{t+\Delta}, w_{t+\Delta}, x_{t+\Delta}) - \varphi(v, w, x)] \\
&= \lim_{\Delta\to 0}\frac{1}{\Delta}E[\varphi(v_{t+\Delta}, w_{t+\Delta}, x_{t+\Delta}) - \varphi(v, w_{t+\Delta}, x_{t+\Delta})] \\
&\quad + \lim_{\Delta\to 0}\frac{1}{\Delta}E\varphi(v, w_{t+\Delta}, x_{t+\Delta}) - \varphi(v, w, x)].
\end{aligned}$$

The first limit defines the generator [5.94] of the evolution v, and the second limit gives the generator of the two-component Markov process

$$\mu^\varepsilon(t), x(t/\varepsilon^2),\ t \geq 0.$$

Since the representation

$$\begin{aligned}
&\lim_{\Delta\to 0}\frac{1}{\Delta}E[\varphi(v, w_{t+\Delta}, x_{t+\Delta}) - \varphi(v, w, x)] \\
&= \lim_{\Delta\to 0}\frac{1}{\Delta}E[\varphi(v, w_{t+\Delta}, x_{t+\Delta}) - \varphi(v, w, x_{t+\Delta})] \\
&= \lim_{\Delta\to 0}\frac{1}{\Delta}E[\varphi(v, w, x_{t+\Delta}) - \varphi(v, w, x)]
\end{aligned}$$

holds true, considering

$$\lim_{\Delta\to 0}\frac{1}{\Delta}E[\varphi(v, w_{t+\Delta}, x_{t+\Delta}) - \varphi(v, w, x_{t+\Delta})] = at^{\frac{1}{2}}\mathbf{\Gamma}^\varepsilon(x)\varphi(v, w, x)$$

and

$$\lim_{\Delta\to 0}\frac{1}{\Delta}E[\varphi(v, w, x_{t+\Delta}) - \varphi(v, w, x)] = \varepsilon^{-2}Q\varphi(v, w, x),$$

we obtain

$$\mathbf{L}^\varepsilon\varphi(v, w, x) = \left[\varepsilon^{-2}\mathbf{Q} + at^{-\frac{1}{2}}\mathbf{\Gamma}^\varepsilon(x) + \mathbf{C}_t^{\nabla,\varepsilon}(x)\right]\varphi(v, w, x).$$

LEMMA 5.23.– *The generator* $\boldsymbol{L}^{\varepsilon}$ *has the asymptotic representation*

$$\boldsymbol{L}^{\varepsilon}\varphi(v,w,x) = (\varepsilon^{-2}\boldsymbol{Q} + \varepsilon^{-1}(at^{-\frac{1}{2}}\boldsymbol{\Gamma}_1(x) + a\boldsymbol{C}_1(x)) + at^{-\frac{1}{2}}\boldsymbol{\Gamma}_2(x) + (at^{-\frac{1}{2}}(v + t^{\frac{1}{2}}w)\boldsymbol{C}_2(x) + \frac{1}{2}t^{-1}v\boldsymbol{D}) + \theta_L^{\varepsilon}(x))\varphi(v,w,x), \quad [5.96]$$

where

$$\boldsymbol{C}_1(x)\varphi(v,w,x) = C'(0,x)\varphi_v'(v,w,x),$$

$$\boldsymbol{C}_2(x)\varphi(v,w,x) = C''(0,x)\varphi_v'(v,w,x),$$

$$\boldsymbol{D}\varphi(v,w,x) = \varphi_v'(v,w,x),$$

and the remainder term $||\theta_L^{\varepsilon}(x)\varphi(v,w,x)|| \to 0$ *at* $\varepsilon \to 0$.

PROOF.– The proof of the lemma follows from the representations of the generators $\mathbf{\Gamma}^{\varepsilon}(x)$ $\mathbf{C}_t^{\nabla,\varepsilon}(x)$.

For the generator [5.96], we consider the truncated operator

$$\boldsymbol{L}_0^{\varepsilon} = \varepsilon^{-2}\mathbf{Q} + \varepsilon^{-1}at^{-\frac{1}{2}}\mathbf{Q}_1(x) + at^{-\frac{1}{2}}\mathbf{Q}_2(x), \quad [5.97]$$

where

$$\mathbf{Q}_1(x) = \mathbf{\Gamma}_1(x) + t^{-\frac{1}{2}}\mathbf{C}_1(x),$$

$$\mathbf{Q}_2(x) = \mathbf{\Gamma}_2(x) + (v + t^{-\frac{1}{2}}w)\mathbf{C}_2(x) + \frac{1}{2at^{\frac{1}{2}}}v\mathbf{D}.$$

LEMMA 5.24.– *The solution of the singular perturbation problem for the truncated operator [5.97] on the test functions*

$$\varphi^{\varepsilon}(v,w,x) = \varphi(v,w) + \varepsilon t^{-\frac{1}{2}}\varphi_1(v,w,x) + \varepsilon^2 t^{-\frac{1}{2}}\varphi_2(v,w,x)$$

under theorem conditions is determined by the relation

$$\mathbf{L}_0^{\varepsilon}\varphi^{\varepsilon}(v,w,x) = t^{-\frac{1}{2}}\mathbf{L}\varphi(v,w) + \varepsilon\theta^{\varepsilon}(x)\varphi(v,w), \quad [5.98]$$

where the remainder term $\theta^{\varepsilon}(x)$ *is uniformly bounded at* x, *and the limit generator* $\mathbf{L}$ *is defined by the relation*

$$\mathbf{L}\Pi = a\Pi\mathbf{Q}_2(x)\Pi + a^2t^{-\frac{1}{2}}\Pi\mathbf{Q}_1(x)R_0\mathbf{Q}_1(x)\Pi.$$

PROOF.– We consider the representation of the generator $\mathbf{L}_0^\varepsilon$ on the test functions $\varphi^\varepsilon(v, w, x)$, i.e.

$$\mathbf{L}_0^\varepsilon \varphi^\varepsilon(v, w, x) = \varepsilon^{-2}\mathbf{Q}\varphi(v, w) + \varepsilon^{-1}t^{-\frac{1}{2}}[\mathbf{Q}\varphi_1(v, w, x) + a\mathbf{Q}_1(x)\varphi(v, w)] \\ + at^{-\frac{1}{2}}\mathbf{Q}_2(x)\varphi(v, w) + at^{-1}\mathbf{Q}_1(x)\varphi_1(v, w, x) \\ + t^{-\frac{1}{2}}\mathbf{Q}\varphi_2(v, w, x) + \varepsilon\theta^\varepsilon(x)\varphi(v, w).$$

Since φ does not depend on x

$$\mathbf{Q}\varphi = 0, \Leftrightarrow \varphi \in N_Q.$$

The balance conditions [5.89] and [5.90] define the solvability of the equation:

$$\mathbf{Q}\varphi_1(v, w, x) + a\mathbf{Q}_1(x)\varphi(v, w) = 0.$$

Therefore,

$$\varphi_1(v, w, x) = aR_0\mathbf{Q}_1(x)\varphi(u, w). \quad [5.99]$$

Consequently, considering [5.99], we obtain the representation

$$t^{-\frac{1}{2}}\mathbf{Q}\varphi_2(v, w, x) + t^{-\frac{1}{2}}\mathbf{L}(x)\varphi(v, w) = t^{-\frac{1}{2}}\mathbf{L}\varphi(v, w),$$

where

$$\mathbf{L}(x)\varphi(v, w) = a\mathbf{Q}_2(x)\varphi(v, w) + a^2t^{-\frac{1}{2}}\mathbf{Q}_1(x)R_0\mathbf{Q}_1\varphi(v, w).$$

For $\varphi_2(v, w, x)$, we have the representation

$$\varphi_2(v, w, x) = R_0[\mathbf{L}(x) - \mathbf{L}]\varphi(v, w). \quad [5.100]$$

Since

$$\Theta^\varepsilon(x)\varphi^\varepsilon(v, w, x) = at^{-1}\mathbf{Q}_1\varphi_2(v, w, x) \\ + at^{-\frac{1}{2}}\mathbf{Q}_2\varphi_1(v, w, x) + at^{-1}\mathbf{Q}_2\varphi_2(v, w, x)$$

and considering [5.99] and [5.100] as well as the smoothness of the function $\varphi(v, w)$, we obtain the statement of the lemma in accordance with the remainder term.

We can now calculate the limit operator $\mathbf{L}$:

$$\mathbf{L}\varphi(v, w) = a\Pi\mathbf{\Gamma}_2(x)\varphi(v, w) + a(v + t^{\frac{1}{2}}w)\Pi\mathbf{C}_2(x)\varphi(v, w) \\ + \frac{1}{2}t^{-\frac{1}{2}}v\mathbf{D}\varphi(v, w) + a^2t^{-\frac{1}{2}}\Pi[\mathbf{\Gamma}_1(x) \\ + t^{\frac{1}{2}}\mathbf{C}_1(x)]R_0[\mathbf{\Gamma}_1(x) + t^{\frac{1}{2}}\mathbf{C}_1(x)]\varphi(v, w)$$

$$= a \int_X \pi(dx) b_2(x) \varphi''_w(v,w)$$

$$+ a(v + t^{\frac{1}{2}} w) \int_X \pi(dx) C''(0,x) \varphi'_v(v,w)$$

$$+ \frac{1}{2} t^{-\frac{1}{2}} v \varphi'_v(v,w) + a^2 t^{-\frac{1}{2}} (\int_X \pi(dx) b_1(x) R_0 b_1(x) \varphi''_w(v,w)$$

$$+ t^{\frac{1}{2}} \int_X \pi(dx) b_1(x) R_0 C'(0,x) \varphi''_{vw}(v,w)$$

$$+ t^{\frac{1}{2}} \int_X \pi(dx) C'(0,x) R_0 b_1(x) \varphi''_{vw}(v,w)$$

$$+ t \int_X \pi(dx) C'(0,x) R_0 C'(0,x) \varphi''_v(v,w)).$$

Therefore,

$$\mathbf{L}\varphi(v,w) = \left[v(ac_2 + \frac{1}{2} t^{-\frac{1}{2}}) + act^{\frac{1}{2}} w \right] \varphi'_v(v,w) + \frac{a}{2} B(t) \varphi''_w(v,w),$$

where

$$B(t)\varphi''(v,w) = B_1(t)\varphi''_w(v,w) + B_2(t)\varphi''_w(v,w) + B_3(t)\varphi''_w(v,w),$$

$$B_1(t) = 2 \int_X \pi(dx) b_2(x) + 2at^{-\frac{1}{2}} \int_X \pi(dx) b_1(x) R_0 b_1(x),$$

$$B_2(t) = 2a \left[\int_X \pi(dx) b_1(x) R_0 C'(0,x) + \int_X \pi(dx) C'(0,x) R_0 b_1(x) \right],$$

$$B_3(t) = 2at^{\frac{1}{2}} \int_X \pi(dx) C'(0,x) R_0 C'(0,x).$$

Considering the Korolyuk model theorem (Korolyuk and Limnios 2005), we complete the proof of the theorem.

We now consider the continuous SOP

$$du^\varepsilon(t) = a(t) C^\varepsilon(u^\varepsilon(t), x(t/\varepsilon^2)) dt, \qquad [5.101]$$

with the regression function

$$C^\varepsilon(u,x) = \nabla_{b(t)} C(u, x(t) + \varepsilon^{-1} C_0(u,x), u \in R, x \in X.$$

The averaged system:

$$\frac{du}{dt} = C'(u), C(u) = \int_X \pi(dx)C(u,x).$$

The balance conditions:

$$\int_X \pi(dx)C_0(u,x) = 0, \int_X \pi(dx)C'(0,x) = 0. \quad [5.102]$$

We denote $C_0^0(x) = C_0(0,x)$, then diffusion perturbation at the extremum point has the form

$$C_0^\varepsilon(t) = \varepsilon^{-1}a\int_{t_0}^{t} C_0(0,x(s/\varepsilon^2))/s^{\frac{1}{2}}ds = \varepsilon^{-1}a\int_{t_0}^{t} C_0^0(x(s/\varepsilon^2)))/s^{\frac{1}{2}}ds.$$

We consider the fluctuations of SOP [5.101] normalized by the relation

$$v^\varepsilon(t) = \varepsilon^{-1}t^{\frac{1}{2}}\left[u^\varepsilon(t) - \varepsilon C_0^\varepsilon(t)\right].$$

THEOREM 5.8.– *We assume that the balance condition [5.102] is fulfilled, as well as the convergence conditions for the SOP. We also assume that the functions* $C_0'(u,x)$ *and* $C_0''(u,x)$ *are uniformly bounded at* x*:*

$$\sup_{x\in X}||C_0'(u,x)|| \le C_1 < +\infty;$$

$$\sup_{x\in X}||C_0''(u,x)|| \le C_2 < +\infty.$$

Then, the weak convergence

$$(v^\varepsilon(t), C_0^\varepsilon(t)) \to (\zeta(t), W_\sigma(t)), t > 0, \qquad \varepsilon \to 0$$

holds true at each finite interval $0 < t_0 < t < T$.

The limit process $(\zeta(t), W_\sigma(t))$ *is given by the generator*

$$\mathbf{L}\varphi(v,w) = \left[v(ac_2 + \frac{1}{2}t^{-\frac{1}{2}}) + ac_2 t^{-\frac{1}{2}}w\right]\varphi_v'(v,w) + \frac{at^{-\frac{1}{2}}}{2}B\varphi_w''(v,w),$$

where

$$c_2 = \int_X \pi(dx)C''(0,x);$$

$$B = 2\int_X \pi(dx)C_0(0,x)R_0C_0(0,x).$$

Furthermore, we require some additional lemmas.

LEMMA 5.25.– *The generator* $\mathbf{L}^\varepsilon$ *has the asymptotic representation of the test functions* $\varphi(v) \in C^4(R)$

$$\mathbf{L}^\varepsilon\varphi(v,w,x) = \Big[\varepsilon^{-2}\mathbf{Q} + \varepsilon^{-1}at^{-\frac{1}{2}}\mathbf{C}^0(x) \\ + \varepsilon^{-1}a\mathbf{C}(x) + at^{-\frac{1}{2}}\mathbf{C}_t^1(x) + \theta_L^\varepsilon(x)\Big]\varphi(v,w,x),$$

where

$$\mathbf{C}^0(x)\varphi(v,w,x) = C_0^0(x)\varphi'_w(v,w,x),$$

$$\mathbf{C}(x)\varphi(v,w,x) = [C(0,x) + zC'_0(0,x)]\,\varphi'_v(v,w,x),$$

$$\mathbf{C}_t^1(x)\varphi(v,w,x) = (t^{\frac{1}{2}}z\,[C'(0,x) + zC''_0(0,x)/2] + \frac{v}{2a}t^{\frac{1}{2}})\varphi'_v(v,w,x),$$

$$z = t^{-\frac{1}{2}}v + w,$$

and the remainder term

$$||\theta_L^\varepsilon(x)\varphi(v,w,x)|| \to 0, \varepsilon \to 0.$$

LEMMA 5.26.– *The solution of the singular perturbation problem for the truncated operator*

$$\mathbf{L}_0^\varepsilon = \varepsilon^{-2}\mathbf{Q} + \varepsilon^{-1}a[t^{-\frac{1}{2}}\mathbf{C}^0(x) + \mathbf{C}(x)] + at^{-\frac{1}{2}}\mathbf{C}_t^1(x),$$

under theorem conditions is determined by the relation

$$\mathbf{L}_0^\varepsilon\varphi^\varepsilon(v,w,x) = t^{-\frac{1}{2}}\mathbf{L}\varphi(v,w) + \varepsilon\theta^\varepsilon(x)\varphi(v,w),$$

where the remainder term $\theta^\varepsilon(x)$ *is uniformly bounded at* x.

The limit operator is defined by the relation

$$\mathbf{L}\Pi = a\Pi\mathbf{C}_t^1(x)\Pi + a^2t^{-\frac{1}{2}}\Pi\mathbf{C}^0(x)R_0\mathbf{C}^0(x)\Pi.$$

PROOF OF THEOREM 5.8.– We can provide it according to the scheme of the proof for Theorem 5.7 using Lemmas 5.25 and 5.26 and by considering the Korolyuk model theorem (Korolyuk and Limnios 2005).

Consequently, the fluctuations of the SOP with impulse perturbations when an equilibrium point exists, at increasing time intervals, are described by the stochastic differential equation. The diffusion component is formed by the first and second moments of the impulse process and from the changing velocity for the regression function at an equilibrium point.

6

Combination of Approximations of Different Types

6.1. Asymptotic properties of a stochastic diffusion process with an equilibrium point of a quality criterion

A random evolution in the form of a diffusion process with a control determined by the condition of reaching the extremum of a quality criterion function was studied in Nikitin (2015) and Nikitin and Khimka (2017). A particular case is the existence of an equilibrium point of the quality criterion used in many applied optimal estimation problems (Nevelson and Khasminsky 1972). The problem of the asymptotic behavior of a system under random disturbances (Korolyuk and Chabanyuk 2002) is considered independently. This problem was investigated using a small parameter in the schemes of series and diffusion approximation (Korolyuk and Korolyuk 1999). To prove important statements, the Korolyuk model theorem (Korolyuk and Limnios 2005) is used.

Let a transfer process $y(t) \in \mathbb{R}^d$ be defined by the stochastic differential equation

$$dy(t) = a(y(t), x(t))dt + \sigma(y(t), x(t), u(t))dW(t), \qquad [6.1]$$

where $x(t), t > 0$ is a uniformly ergodic Markov process defined by the generator

$$\mathbf{Q}\varphi(x) = q(x) \int_X P(x, dy)[\varphi(y) - \varphi(x)] \qquad [6.2]$$

in a measurable phase space $(X, \mathbf{X})$ on the Banach space $B(X)$ of real-valued bounded functions $\varphi(x)$ with the supremum norm

$$\|\varphi\| = \sup_{x \in X} |\varphi(x)|.$$

The generator $\mathbf{Q}$ is reducibly inverse on $B(X)$ with the projector

$$\Pi\varphi(x) := \int\limits_X \pi(dx)\varphi(x),$$

where $\pi(B)$, $(B \in X)$ is the stationary distribution of the Markov process $x(t)$, $t \geq 0$; this distribution is found from the relationships

$$\pi(dx)q(x) = q\rho(dx), \; q = \int\limits_X \pi(dx)q(x),$$

where $p(dx)$ is the stationary distribution of the embedded Markov chain x_n, $n \geq 0$ and the potential R_0 of the Markov semigroup

$$R_0 = \Pi - (\Pi + \mathbf{Q})^{-1}.$$

The functions $a(y,x) = (a_k(y,x), k = \overline{1,d})$ and $\sigma(y,x,u) = (\sigma_k(y,x,u), k = \overline{1,d})$, $y \in \mathbf{R}^d$, $x \in X$, satisfy the conditions of the existing global solution for the evolution equation

$$dy_x(t) = a(y_x(t),x)dt + \sigma(y_x(t),x,u_x(t))dW(t), \; x \in X \qquad [6.3]$$

for each fixed value of x of the Markov process $x(t)$, $t \geq 0$, in an interval $[\tau_i, \tau_{i+1}]$ of stay of the process $x(t)$, $t \geq 0$, at a state $x \in X$.

Let the quality criterion of the transfer process [6.1] be defined by a function $G(y,x,u)$, $y \in \mathbf{R}^d$, with only one equilibrium point u_x^* in the interval $[\tau_i, \tau_{i+1}]$, that follows from the condition $G(y_x, x, u_x) = 0$, or, in the general representation [6.1], the control $u(t)$ is defined by the condition

$$G(y(t),x(t),u(t)) = 0. \qquad [6.4]$$

Note that the solution of stochastic equation [6.1] in the interval $[\tau_i, \tau_{i+1}]$ under the nonrandom control $u(t)$ forms a Markov process.

To find the asymptotic properties of the solution of problem [6.1], [6.4] in a scheme of series with a small parameter $\varepsilon > 0$, we consider the stochastic equation

$$dy^\varepsilon(t) = a(y^\varepsilon(t),x(t/\varepsilon))dt + \sigma(y^\varepsilon(t),x(t/\varepsilon),u^\varepsilon(t))dW(t), \qquad [6.5]$$

and the stochastic approximation procedure

$$du^\varepsilon(t) = \alpha(t)G(y^\varepsilon(t),x(t/\varepsilon),u^\varepsilon(t))dt \qquad [6.6]$$

with the common initial conditions

$$x(0) = x_0, \; y(0) = y_0, \; u(0) = u_0. \qquad [6.7]$$

THEOREM 6.1.– *Let* $a(y,x) \in C(\mathbf{R}^d, X)$, $\sigma(y,x,u) \in C(\mathbf{R}^d, X, \mathbf{R}^d)$ *and* $G(y,x,u) \in C(\mathbf{R}^d, X, \mathbf{R}^d)$

Then, $\varepsilon < \varepsilon_0$ *is sufficiently small, and the following weak convergence takes place:*

$$(y^\varepsilon(t), u^\varepsilon(t)) \Rightarrow (\hat{y}(t), \hat{u}(t)), \qquad [6.8]$$

where the limit process $(\hat{y}^\varepsilon(t), \hat{u}^\varepsilon(t))$ *is specified by the generator*

$$L\varphi(y,u) = A(y,u)\varphi(y,u) + \frac{1}{2}B(y,u)\varphi(y,u) \qquad [6.9]$$

with the following representation in terms of the test functions $\varphi(y,u) \in C^{3,2}(\mathbb{R}^d, \mathbb{R}^d)$*:*

$$A(y,u)\varphi(y,u) = a(y)\varphi'_y(y,u) + \alpha(t)G(y,u)\varphi'_u(y,u), \qquad [6.10]$$

where $a(y) = \int_X a(y,x)\pi(dx)$, $G(y,u) = \int_X G(y,x,u)\pi(dx)$, $B(y,u)\varphi(y,u) = \hat{\sigma}^2(y,u)\varphi''_{yy}(y,u)$, $\hat{\sigma}^2(y,u) = \int_X \sigma^2(y,x,u)\pi(dx)$.

COROLLARY 6.1.– *We will describe the limit control process* $(\hat{y}(t), \hat{u}(t))$ *by the equations*

$$d\hat{y}(t) = a(\hat{y}(t))dt + \sigma(\hat{y}(t), \hat{u}(t))dW(t), \qquad [6.11]$$

$$d\hat{u}(t) = \alpha(t)G(\hat{y}(t), \hat{u}(t))dt. \qquad [6.12]$$

COROLLARY 6.2.– *Let us consider the transfer process described in a scheme of series by the stochastic differential equation*

$$dy^\varepsilon(t) = a(y^\varepsilon(t), x(t/\varepsilon), u^\varepsilon(t))dt + \sigma(y^\varepsilon(t), x(t/\varepsilon), u^\varepsilon(t))dW(t) \qquad [6.13]$$

with the control $u^\varepsilon(t)$ *defined by the equation*

$$du^\varepsilon(t) = \alpha(t)G(y^\varepsilon(t), x(t/\varepsilon), u^\varepsilon(t))dt$$

and components $a(y,x,u)$, $G(y,x,u)$, $\sigma(y,x,u) \in C(\mathbf{R}^d, X, \mathbf{R}^d)$.

Then, the weak convergence

$$(y^\varepsilon(t), u^\varepsilon(t)) \Rightarrow (\hat{y}(t), \hat{u}(t))$$

takes place, where the limit process is defined over the test functions $\varphi(y,x,u) \in C^{3,0,3}(\mathbf{R}^d, X, \mathbf{R}^d)$ by the generator [6.9], where

$$A(y,u)\varphi(y,u) = a(y,u)\varphi'_y(y,x) + G(y,u)\varphi'_u(y,u),$$

$$a(y,u) = \int\limits_X a(y,x,u)\pi(dx).$$

We first establish some properties of the generator of the three-component Markov process $y_t^\varepsilon = y^\varepsilon(t)$, $x_t^\varepsilon = x^\varepsilon(t)$, $u_t^\varepsilon = u^\varepsilon(t)$, defined by the relationship

$$L^\varepsilon(y,x)\varphi(y,x,u) = \lim_{\Delta\to\infty}\frac{1}{\Delta}\mathbf{E}[\varphi(y_{t+\Delta t}^\varepsilon, x_{t+\Delta t}^\varepsilon, u_{t+\Delta t}^\varepsilon) \\ - \varphi(y,x,u)|y_t^\varepsilon = y; x_t^\varepsilon = x, u_t^\varepsilon = u].$$

We introduce the following denotations for the conditional expectation with the corresponding decompositions of increments:

$$\mathbf{E}_{y,x,u}\varphi(y+\Delta y, x_{t+\Delta t}^\varepsilon, u+\Delta u) \\ = \mathbf{E}[\varphi(y+\Delta y, x_{t+\Delta t}^\varepsilon, u+\Delta u)|y_t^\varepsilon = y, x_t^\varepsilon = x, u_t^\varepsilon = u].$$

Since

$$\mathbf{E}_{y,x,u}\varphi(y+\Delta y, x_{t+\Delta t}^\varepsilon, u+\Delta u) \\ = \mathbf{E}_{y,x,u}\varphi\left(y+\int\limits_t^{t+\Delta t} a(y^\varepsilon(s),x)ds + \int\limits_t^{t+\Delta t}\sigma(y^\varepsilon(s),x,u^\varepsilon(s))dW(s), x, u+\Delta u\right) \\ \times I(\theta > \varepsilon^{-1}\Delta) + \mathbf{E}_{y,x,u}\varphi(y+\int\limits_t^{t+\Delta t} a(y^\varepsilon(s), x_{t+\Delta t}^\varepsilon)ds \\ + \int\limits_t^{t+\Delta t}\sigma(y^\varepsilon(s), x_{t+\Delta}^\varepsilon, u^\varepsilon(s))dW(s), x_{t+\Delta t}^\varepsilon, u+\Delta u)I(\theta \le \varepsilon^{-1}\Delta) + o(\Delta), \tag{6.14}$$

where θ is the sojourn time of the Markov process $x(t)$, $t \ge 0$, at a state x, we have

$$I(\theta > \varepsilon^{-1}\Delta) = 1 - \varepsilon^{-1}q(x)\Delta + o(\Delta), \\ I(\theta \le \varepsilon^{-1}\Delta) = \varepsilon^{-1}q(x)\Delta + o(\Delta).$$

For the first addend in sum [6.14], we have

$$\varphi(y+\int\limits_t^{t+\Delta t} a(y^\varepsilon(s),x)ds + \int\limits_t^{t+\Delta t}\sigma(y^\varepsilon(s),x,u^\varepsilon(s))dW(s), x, u+\Delta u) \\ = \varphi(v+\int\limits_t^{t+\Delta}\sigma(y^\varepsilon(s),x,u^\varepsilon(s))dW(s), x, u+\Delta u),$$

where $v = y + \int\limits_t^{t+\Delta t} a(y^\varepsilon(s),x)ds$.

For the latter representation of the first function with allowance for $\pm\varphi(v, x, u + \Delta u)$

$$\varphi(v + \int\limits_t^{t+\Delta t} \sigma(y^\varepsilon(s), x, u^\varepsilon(s))dW(s), x, u + \Delta u)$$

$$= \varphi'_y(v, x, u + \Delta u) \int\limits_t^{t+\Delta t} \sigma(y^\varepsilon(s), x, u^\varepsilon(s))dW(s)$$

$$+\frac{1}{2}\varphi''_{yy}(v, x, u+\Delta u)[\int\limits_t^{t+\Delta t} \sigma(y^\varepsilon(s), x, u^\varepsilon(s))dW(s)]^2+\varphi(v, x, u+\Delta u)+o(\Delta).$$

[6.15]

Since

$$\varphi'_y(v, x, u + \Delta u) = \varphi'_y(v, x, u) + \varphi''_{uy}(v, x, u)\Delta u + o(\Delta) =$$
$$= \varphi'_y(v, x, u) + \varphi''_{uy}(v, x, u)\alpha(t)G(y, x, u)u + o(\Delta),$$
$$\varphi''_{yy}(v, x, u + \Delta u) = \varphi''_{yy}(v, x, u) + \varphi'''_{yyu}(v, x, u)\alpha(t)G(y, x, u)\Delta u + o(\Delta),$$

for the function [6.15], we obtain

$$\varphi(v + \int\limits_t^{t+\Delta t} \sigma(y^\varepsilon(s), x, u^\varepsilon(s))dW(s), x, u + \Delta u)$$

$$= \varphi(v, x, u) + \varphi'_u(v, x, u)\alpha(t)G(y, x, u)\Delta + o(\Delta)$$

$$+\varphi'_y(v, x, u) \int\limits_t^{t+\Delta t} \sigma(y^\varepsilon(s), x, u^\varepsilon(s))dW(s)$$

$$+\alpha(t)\varphi''_{yu}(v, x, u)G(y, x, u) \int\limits_t^{t+\Delta} \sigma(y^\varepsilon(s), x, u^\varepsilon(s))dW(s)\Delta + o(\Delta)$$

$$+\frac{1}{2}\varphi''_{yy}(v, x, u)[\int\limits_t^{t+\Delta} \sigma(y^\varepsilon(s), x, u^\varepsilon(s))dW(s)]^2+$$

$$+\frac{1}{2}\alpha(t)\varphi'''_{yyu}(v, x, u)[\int\limits_t^{t+\Delta} \sigma(y^\varepsilon(s), x, u^\varepsilon(s))dW(s)]^2G(y, x, u)\Delta + o(\Delta).$$

[6.16]

Taking into account the representation of the variable v and continuous differentiability of the test functions φ, we obtain

$$\varphi(v,x,u) = \varphi(y + \int_t^{t+\Delta} a(y^\varepsilon(s),x)ds, x)$$

$$= \varphi(y,x,u) + \varphi'_y(y,x,u)a(y,x)\Delta + o(\Delta).$$

All the components with the variable v have similar representations in representation [6.16]. Therefore, according to representation [6.16], we obtain

$$\varphi(v + \int_t^{t+\Delta} \sigma(y^\varepsilon(s),x,u^\varepsilon(s))dW(s), x, u+\Delta u)$$

$$= \varphi(y,x,u) + \varphi'_y(y,x,u)a(y,x)\Delta + o(\Delta) + \alpha(t)\varphi'_u(y,x,u)G(y,x,u)\Delta$$

$$+ o(\Delta) + \varphi'_y(y,x,u)a(y,x)\int_t^{t+\Delta} \sigma(y^\varepsilon(s),x,u^\varepsilon(s))dW(s)\Delta + o(\Delta)$$

$$+ \alpha(t)\varphi''_{yu}(y,x,u)G(y,x,u)\int_t^{t+\Delta} \sigma(y^\varepsilon(s),x,u^\varepsilon(s))dW(s)\Delta + o(\Delta)$$

$$+ \frac{1}{2}\varphi''_{yy}(v,x,u)[\int_t^{t+\Delta} \sigma(y^\varepsilon(s),x,u^\varepsilon(s))dW(s)]^2 + o(\Delta)$$

$$+ \frac{1}{2}\alpha(t)\varphi'''_{yyu}(v,x,u)[\int_t^{t+\Delta} \sigma(y^\varepsilon(s),x,u^\varepsilon(s))dW(s)]^2 G(y,x,u)\Delta + o(\Delta).$$

Since the following relationships hold for the conditional expectation:

$$\mathbf{E}_{u,x,y}\int_t^{t+\Delta} \sigma(y^\varepsilon(s),x,u^\varepsilon(s))dW(s) = 0,$$

$$\mathbf{E}_{u,x,y}[\int_t^{t+\Delta} \sigma(y^\varepsilon(s),x,u^\varepsilon(s))dW(s)] = \sigma^2(y,x,u)\Delta + o(\Delta),$$

we obtain

$$\begin{aligned}
&\mathbb{E}_{u,x,y}[\varphi(y+\Delta y, x^{\varepsilon}_{t+\Delta}, u+\Delta u)]\\
&=\varphi(y,x,u)+[\varphi'_y(y,x,u)a(y,x)+\alpha(t)\varphi'_u(y,x,u)G(y,x,u)]\Delta\\
&+\frac{1}{2}\varphi''_{yy}(y,x,u)\sigma^2(y,x,u)\Delta-\varepsilon^{-1}q(x)\mathbb{E}_{y,x,u}\varphi(y,x,u)\Delta+\\
&+\varepsilon^{-1}q(x)\mathbb{E}_{y,x,u}\varphi(y,x^{\varepsilon}_{t+\Delta},u)\Delta+o(\Delta).
\end{aligned}$$

Thus, for the generator $L^{\varepsilon}(y,x)$, we have

$$\begin{aligned}
L^{\varepsilon}(y,x)\varphi(y,x,u) &= \lim_{\Delta\to 0}\frac{1}{\Delta}\varepsilon^{-1}q(x)\mathbb{E}_{y,x,u}[\varphi(y,x^{\varepsilon}_{t+\Delta},u)-\varphi(y,x,u)]\\
&+\varphi'_y(y,x,u)a(y,x)+\alpha(t)\varphi'(x)(y,x,u)G(y,x,u)\\
&+\frac{1}{2}\varphi''_{yy}(y,x,u)\sigma^2(y,x,u)\\
&=\varepsilon^{-1}Q\varphi(y,x,u)+\varphi'_y(y,x,u)a(y,x)\\
&+\alpha(t)\varphi'_u(x)(y,x,u)G(y,x,u)\\
&+\frac{1}{2}\varphi''_{yy}(y,x,u)\sigma^2(y,x,u).
\end{aligned}$$

We will formulate the above reasoning in the form of the following statement.

LEMMA 6.1.– *The representation of the generator of the three-component Markov process*

$$y^{\varepsilon}_t := y^{\varepsilon}(t),\ x^{\varepsilon}_t := x(t/\varepsilon),\ u^{\varepsilon}_t := u^{\varepsilon}(t),\ t\geq 0,$$

in terms of the test functions $\varphi(y,x,u)\in C^{3,0,2}(\mathbb{R}^d,X,\mathbb{R}^d)$ *is as follows:*

$$L^{\varepsilon}(y,x)\varphi(y,x,u)=\varepsilon^{-1}Q\varphi(y,x,u)+L(x)\varphi(y,x,u), \qquad [6.17]$$

$$\begin{aligned}
&L(x)\varphi(y,x,u)=\varphi'_y(y,x,u)a(y,x)+\alpha(t)\varphi'_u(y,x,u)G(y,x,u)+\\
&+\frac{1}{2}\varphi''_{yy}(y,x,u)\sigma^2(y,x,u).
\end{aligned}$$

LEMMA 6.2.– *The solution of the singular perturbation problem for the generator [6.17] in terms of the test functions* $\varphi^{\varepsilon}(y,x,u)=\varphi(y,u)+\varepsilon\varphi_1(y,x,u)$ *defines the limit generator*

$$L\varphi(y,u)=L_y\varphi(y,u)+L_u\varphi(y,u),$$

where

$$L_y\varphi(y,u)=a(y)\varphi'_y(y,u)+\frac{1}{2}\sigma^2(y,u)\varphi''_{yy},$$

$$
\begin{aligned}
&L_u\varphi(y,u) = \alpha(t)G(y,u)\varphi'_u(y,u),\\
&a(y) = \int_X a(y,x)\pi(dx),\\
&G(y,u) = \int_X G(y,x,u)\pi(dx),\\
&\sigma^2(y,u) = \int_X \sigma^2(y,x,u)\pi(dx)
\end{aligned}
$$

PROOF.– Consider the representation

$$
\begin{aligned}
L^\varepsilon(y,x)\varphi^\varepsilon(y,x,u) = \varepsilon^{-1}\mathbf{Q}\varphi(y,u) + \mathbf{Q}\varphi_1(y,x,u)\\
+ L(x)\varphi(y,u) + \varepsilon L(x)\varphi_1(y,x,u),
\end{aligned}
$$

where

$$
\begin{aligned}
L(x) = \varphi'_y(y,u)a(y,x) + \alpha(t)G(y,x,u)\varphi'_u(y,u)\\
+ \frac{1}{2}\sigma^2(y,x,u)\varphi''_{yy}(y,u) + \varepsilon\theta(x),
\end{aligned}
$$

with the remainder in the form $\theta(x) = L(x)\varphi_1(y,x,u) = \theta_y(x) + \theta_u(x)$.

We write the expression $\varphi_1(y,x,u)$ in the form

$$
\varphi_1(y,x,u) = R[L - L(x)]\varphi(y,u) = R_0\tilde{L}(x)\varphi(y,u),
$$

where

$$
\tilde{L}(x) = \tilde{a}(y,x)\varphi'_y(y,u) + \alpha(t)\tilde{G}(y,x,u)\varphi'_u(y,u) + \frac{1}{2}\tilde{\sigma}^2(y,x,u)\varphi''_{yy}(y,u),
$$

$$
\begin{aligned}
&\tilde{a}(y,x) = a(y) - a(y,x),\\
&\tilde{G}(y,x,u) = G(y,u) - G(y,x,u),\\
&\tilde{\sigma}^2(y,x,u) = \sigma^2(y,u) - \sigma^2(y,x,u).
\end{aligned}
$$

Thus, for the remainder terms $\theta_y(x)$ and $\theta_u(x)$, we have

$$
\begin{aligned}
\theta_y(x) = a(y,x)R_0[\tilde{a}(y,x)\varphi'_y(y,u)]'_y + \frac{1}{2}a(y,x)R_0[\tilde{\sigma}^2(y,x,u)\varphi''_{yy}]'_y\\
+\frac{1}{2}\sigma^2(y,x,u)R_0[\tilde{a}(y,x)\varphi'_y(y,u)]''_{yu}\\
+\frac{1}{4}\sigma^2(y,x,u)R_0[\tilde{\sigma}^2(y,x,u)\varphi''_{yy}(y,u)]''_{yu},
\end{aligned}
$$

$$\begin{aligned}\theta_u(x) &= \alpha(t)a(y,x)R_0[\tilde{G}(y,x,u)\varphi'_u(y,u)]'_u \\ &+\alpha(t)G(y,x,u)R_0[\tilde{a}(y,x)\varphi'_y(y,u)]'_u \\ &+\frac{1}{2}\alpha(t)G(y,x,u)R_0[\tilde{\sigma}^2(y,x,u)\varphi''_{yu}(y,u)]'_u \\ &+\frac{1}{2}\alpha(t)\sigma^2(y,x,u)R_0[\tilde{G}(y,x,u)\varphi'_u(y,u)]''_{yy} \\ &+\alpha^2(t)G(y,x,u)R_0[\tilde{G}(y,x,u)\varphi'_u(y,u)]'_u.\end{aligned}$$

By the Korolyuk theorem (Korolyuk and Korolyuk 1999), $L_y\varphi(y,u)$ defines a limit diffusion process that satisfies the equation

$$d\hat{y}(t) = a(\hat{y})dt + \sigma(\hat{y},\hat{u})dW(t)$$

with the following control:

$$d\hat{u} = \alpha(t)G(\hat{y},\hat{u}(t))dt.$$

PROOF OF THEOREM 6.1.– The statement of Theorem 6.3 follows from the Korolyuk model theorem (Korolyuk and Limnios 2005) and the result of Lemma 6.2.

THEOREM 6.2.– *Let the Lyapunov function $V(y,u)$ of the averaged system*

$$\frac{du}{dt} = G(y,u)$$

be such that it satisfies the following conditions:

$$\begin{aligned}&Y1:\ G(y,u)V'(y,u) < -cV(y,u), \\ &Y2:\ |a(y,x)R_0[\tilde{G}(y,x,u)V'_u(y,u)]'_y| \leq c_1V(y,u), \\ &|G(y,x,u)R_0[\tilde{a}(y,x)V'_y(y,u)]'_u| \leq c_2V(y,u), \\ &|G(y,x,u)R_0[\tilde{\sigma}^2(y,x,u)V''_{yu}(y,u)]'_u| \leq c_3V(y,u), \\ &|\sigma^2(y,x,u)R_0[\tilde{G}(y,x,u)V'_u(y,u)]''_{yy}| \leq c_4V(y,u), \\ &|G(y,x,u)R_0[\tilde{G}(y,x,u)V'_u(y,u)]'_u| \leq c_5(1+V(y,u)).\end{aligned}$$

Next, let the function $\alpha(t)$ be such that

$$\int_0^\infty \alpha(t)dt = \infty, \quad \int_0^\infty \alpha^2(t)dt < \infty.$$

Then, for $\forall\varepsilon > 0$ and $\varepsilon \leq \varepsilon_0$, where ε_0 is sufficiently small, the convergence

$$P\lim_{t\to\infty} u(t) = u^* = 1$$

takes place.

PROOF.– Consider now the generator of the limit control

$$L_u^\varepsilon V(y,u) = L_u V(y,u) + \varepsilon \theta_u(x),$$

for which we obtain the following estimate from conditions $Y1$ and $Y2$:

$$L_u^\varepsilon V(y,u) \leq -c\alpha(t)V(y,u) + c^*\alpha^2(t)(1+V(y,u)),$$

which implies the statement of Theorem 6.2.

The asymptotic value of the control u^* allows us to consider fluctuations of a derivation of the control $u(t)$ from u^*, as well as to establish its basic characteristics.

Furthermore, we consider a transfer process with Markov perturbations and control over the conditions of existence of an equilibrium point of the quality criterion with Markov switchings (Nikitin 2015). For this control, we construct a normalized process and establish its asymptotic normality in the form of an Ornstein–Uhlenbeck process.

We consider the case where the transfer process, i.e. a random evolution, varies under the influence of Markov switchings along the trajectory of new evolution from the state in which it was at the time of switching (regarded as the initial state)

Assume that the transfer process $y^\varepsilon(t) \in \mathbf{R}^d$ is determined by the following stochastic differential equation:

$$dy^\varepsilon(t) = a(y^\varepsilon(t), x(t/\varepsilon^2))dt + \sigma(y^\varepsilon(t), x(t/\varepsilon^2), u^\varepsilon(t))dW(t), \qquad [6.18]$$

where $x(t), t > 0$ is a uniformly ergodic Markov process in a measurable phase space of states $(X, \mathbf{X})$, specified by the generator

$$\mathbf{Q}\varphi(x) = q(x)\int_X P(x,dy)[\varphi(y) - \varphi(x)] \qquad [6.19]$$

in a Banach space $B(X)$ of real-valued bounded functions $\varphi(x)$ with the supremum norm

$$\|\varphi\| = \sup_{x\in X} |\varphi(x)|.$$

The generator $\mathbf{Q}$ is reducibly invertible on $B(X)$ with the projector

$$\Pi\varphi(x) := \int_X \pi(dx)\varphi(x),$$

where $\pi(B)$, $(B \in X)$, is a stationary distribution of the Markov process $x(t)$, $t \geq 0$, given by the formula

$$\pi(dx)q(x) = q\rho(dx), \;\; q = \int_X \pi(dx)q(x)$$

and $\rho(dx)$ is a stationary distribution of the embedded Markov chain x_n, $n \geq 0$ and by the potential R_0 with the following operator representation:

$$R_0 = \Pi - (\Pi + \mathbf{Q})^{-1}.$$

The functions $a(y,x) = (a_k(y,x), k = \overline{1,d})$, $\sigma(y,x,u) = (\sigma_k(y,x,u), k = \overline{1,d})$, $y \in \mathbf{R}^d$, $x \in X$, satisfy the conditions of the existing global solution for the evolutionary equations

$$dy_x(t) = a(y_x(t), x(t))dt + \sigma(y_x(t), x(t), u_x(t))dW(t), \ \ x \in X$$

for every fixed value of x of the Markov process $x(t)$, $t \geq 0$, on the segment $[\tau_i, \tau_{i+1}]$ of stay of the process $x(t)$, $t \geq 0$, in the state $x \in X$.

Assume that the control $u(t)$ in the general representation [6.18] is specified by the condition

$$du^\varepsilon(t) = \alpha(t)G(y^\varepsilon(t), x(t/\varepsilon^2), u^\varepsilon(t)) = 0, \tag{6.20}$$

where the conditions imposed on the function $\alpha(t)$ take the form

$$\int_0^\infty \alpha(t)dt = \infty, \ \int_0 \infty\alpha^2(t)dt < \infty. \tag{6.21}$$

In particular, conditions [6.21] are satisfied for

$$\alpha(t) = \frac{\alpha}{t},$$

which is considered in the following.

Note that condition [6.21] guarantees the convergence of the control $u^\varepsilon(t)$ to the equilibrium point of the quality criterion, i.e. to the point u^*, determined from the condition

$$G(y,u^*) = 0,$$

$$G(y,u) = \int_X \pi(dx)G(y,x,u).$$

The normalized control has the form

$$\nu^\varepsilon(t) = \frac{\sqrt{t}}{\varepsilon}u^\varepsilon(t), \tag{6.22}$$

and the balance condition takes the form

$$\Pi G(y,x,0) = \int_X \pi(dx)G(y,x,0) = 0. \tag{6.22*}$$

THEOREM 6.3.– *Under the conditions of convergence [6.21] for problem [6.18], [6.19] and the additional conditions*

$$(D1):\ \sigma_v^2(y) = 2\int_X \pi(dx)G(y,x,0)\mathbf{R}_0 G(y,x,0) > 0$$

$$(D2):\ \alpha g(y) < -\frac{1}{2},\ g(y) = \int_X \pi(dx)G'_v(y,x,0)$$

the weak convergence

$$v^\varepsilon(t) \Rightarrow \zeta(t),\ \varepsilon \to 0$$

is realized in every finite interval $(0 < t_0 < t < T)$.

The limit process $\zeta(t),\ t > 0$ *is an Ornstein–Uhlenbeck process specified by the generator*

$$\mathbf{L}_v\varphi(y,v) = v(\alpha g(y) + \frac{1}{2})\varphi'_v(y,v) + \frac{1}{2}\alpha^2\hat{\sigma}^2(y)\varphi''_{vv}(y,v).$$

We now establish several auxiliary properties of the transfer process $y^\varepsilon(t)$ and the normalized control $v^\varepsilon(t)$.

LEMMA 6.3.– *The processes* $y^\varepsilon(t)$ *and* $v^\varepsilon(t)$ *are solutions of the stochastic differential equations*

$$dy^\varepsilon(t) = a(y^\varepsilon(t), x_t^\varepsilon)dt + \sigma(y^\varepsilon(t), x_t^\varepsilon, \frac{\varepsilon}{\sqrt{t}}v^\varepsilon(t))dW(t), \qquad [6.23]$$

$$dv^\varepsilon(t) = \varepsilon^{-1}\frac{\alpha}{\sqrt{t}}G(y^\varepsilon(t), x_t^\varepsilon, \frac{\varepsilon}{\sqrt{t}}v^\varepsilon(t))dt + \frac{\nu^\varepsilon(t)}{2t}dt, \qquad [6.24]$$

where

$$x_t^\varepsilon = x(t/\varepsilon^2).$$

PROOF.– Using relation [6.22], we obtain

$$u^\varepsilon(t) = \frac{\varepsilon}{\sqrt{t}}v^\varepsilon(t).$$

Thus, from [6.19], we obtain [6.23] and [6.24].

LEMMA 6.4.– *The generator of a three-component Markov process*

$$y_t^\varepsilon := y^\varepsilon(t),\ x_t^\varepsilon := x(t/\varepsilon^2),\ u_t^\varepsilon := u^\varepsilon(t),\ t \geq 0, \qquad [6.25]$$

has the form

$$\mathbf{L}_t^\varepsilon(x)\varphi(y,x,v) = \varepsilon^{-2}\mathbf{Q}_t^\varepsilon(x)\varphi(y,x,v) + \mathbf{V}_t^\varepsilon(x)\varphi(y,x,v), \qquad [6.26]$$

where

$$\mathbf{V}_t^\varepsilon(x)\varphi(y,x,v) = [\varepsilon^{-1}\frac{\alpha}{\sqrt{t}}G(y,x,\frac{\varepsilon}{\sqrt{t}v}) + \frac{v}{2t}]\varphi_v'(y,x,v)$$
$$+ \varphi_y'(y,x,v)a(y,x) + \frac{1}{2}\varphi_{yy}''(y,x,v)\sigma^2(y,x,\frac{\varepsilon}{\sqrt{t}}v).$$

PROOF.– To construct the generator of process [6.25], we determine the conditional expectation

$$\mathbb{E}[\varphi(y_{t+\Delta}^\varepsilon, x_{t+\Delta}^\varepsilon, v_{t+\Delta}^\varepsilon) - \varphi(y,x,v)|_{y^\varepsilon(t)=y,x^\varepsilon(t)=x,v^\varepsilon(t)=v}]$$
$$= \mathbb{E}[\varphi(y+\Delta y^\varepsilon, x, v+\Delta v^\varepsilon) - \varphi(y,x,v)]I(\theta > \varepsilon^{-2}\Delta)$$
$$+\mathbb{E}[\varphi(y+\Delta y^\varepsilon, x_{t+\Delta}^\varepsilon, v+\Delta v^\varepsilon) - \varphi(y,x,v)]I(\theta < \varepsilon^{-2}\Delta), \qquad [6.27]$$

where θ is the time of stay of the Markov process $x(t),\ t > 0$, in the state x.

Furthermore, we take into account the representations

$$I(\theta \geq \varepsilon^{-2}\Delta) = 1 - \varepsilon^{-2}q(x)\Delta + o(\Delta),$$
$$I(\theta < \varepsilon^{-2}\Delta) = \varepsilon^{-2}q(x)\Delta + o(\Delta).$$

It follows from [6.23] that

$$y_{t+\Delta}^\varepsilon = y + \Delta y^\varepsilon = y + \int_t^{t+\Delta} a(y^\varepsilon(s), x_s^\varepsilon)ds$$
$$+ \int_t^{t+\Delta} \sigma(y^\varepsilon(s), x_s^\varepsilon, \frac{\varepsilon}{\sqrt{t}}v^\varepsilon(s))dW(s) = \bar{y}$$
$$+ \int_t^{t+\Delta} \sigma(y^\varepsilon(s), x_s^\varepsilon, \frac{\varepsilon}{\sqrt{t}}v^\varepsilon(s))dW(s),$$

where

$$\bar{y} = y + \int_t^{t+\Delta} a(y^\varepsilon(s), x_s^\varepsilon)ds.$$

Consider

$$\varphi(y+\Delta y^{\varepsilon},x,v+\Delta v^{\varepsilon})=\varphi(\bar{y}+\int\limits_{t}^{t+\Delta}\sigma(y^{\varepsilon}(s),x_s^{\varepsilon},\frac{\varepsilon}{\sqrt{s}})v^{\varepsilon}(s))dW(s),x,v+\Delta v^{\varepsilon})$$

$$=\varphi(\bar{y},x,v+\Delta v^{\varepsilon})+\varphi_y'(\bar{y},x,v+\Delta v^{\varepsilon})\int\limits_{t}^{t+\Delta}\sigma(y^{\varepsilon}(s),x_s^{\varepsilon},\frac{\varepsilon}{\sqrt{s}})v^{\varepsilon}(s))dW(s)$$

$$+\frac{1}{2}\varphi_{yy}''(\bar{y},x,v+\Delta v^{\varepsilon})[\int\limits_{t}^{t+\Delta}\sigma(y^{\varepsilon}(s),x_s^{\varepsilon},\frac{\varepsilon}{\sqrt{s}})v^{\varepsilon}(s))dW(s)]^2+o(\Delta).$$

From [6.24], we find

$$\varphi_y'(\bar{y},x,v+\Delta v^{\varepsilon})=\varphi_y'(\bar{y},x,v)+\varphi_{yy}''(\bar{y},x,v)[\varepsilon^{-1}\frac{\alpha}{\sqrt{t}}G(y,x,v)+\frac{v}{t}]\Delta+o(\Delta),$$

and

$$\begin{aligned}\varphi(\bar{y},x,v+\Delta v^{\varepsilon})&=\varphi(y,x,v+\Delta v^{\varepsilon})\\&\quad+\varphi_y'(y,x,v+\Delta v^{\varepsilon})a(y,x)\Delta+o(\Delta)\\&=\varphi(y,x,v)+\varphi_v'(y,x,v)[\varepsilon^{-1}\frac{\alpha}{\sqrt{t}}G(y,x,v)+\frac{v}{t}]\Delta\\&\quad+\varphi_y'(y,x,v)a(y,x)\Delta+o(\Delta).\end{aligned}$$

Similarly, we obtain

$$\begin{aligned}\varphi_y'(\bar{y},x,v+\Delta v^{\varepsilon})&=\varphi_y'(y,x,v)\\&\quad+\varphi_{yv}''(y,x,v)[\varepsilon^{-1}\frac{\alpha}{\sqrt{t}}G(y,x,v)+\frac{v}{t}]\Delta\\&\quad+\varphi_{yy}''(y,x,v)a(y,x)\Delta+o(\Delta),\end{aligned}$$

and

$$\begin{aligned}\varphi_{yv}''(\bar{y},x,v+\Delta v^{\varepsilon})&=\varphi_{yy}''(y,x,v)\\&\quad+\varphi_{yyv}'''(y,x,v)[\varepsilon^{-1}\frac{\alpha}{\sqrt{t}}G(y,x,\frac{\varepsilon}{\sqrt{t}}u)+\frac{v}{t}]\Delta\\&\quad+\varphi_{yyyy}^{IV}(y,x,v)a(y,x)\Delta+o(\Delta).\end{aligned}$$

This yields

$$\varphi(y+\Delta y^{\varepsilon}, x, v+\Delta v^{\varepsilon}) = \varphi(y,x,v)+\varphi_v'(y,x,v)[\varepsilon^{-1}\frac{\alpha}{\sqrt{t}}G(y,x,\frac{\varepsilon}{\sqrt{t}}v)+\frac{v}{2t}]\Delta$$

$$+\varphi_y'(y,x,v)a(y,x)\Delta$$

$$+[\varphi_y'(y,x,v+\varphi_{yv}''(y,x,v)[\varepsilon^{-1}\frac{\alpha}{\sqrt{t}}G(y,x,\frac{\varepsilon}{\sqrt{t}}v)+\frac{v}{2t}]\Delta$$

$$+\varphi_{yv}''(y,x,v)a(y,x)\Delta]\int\limits_{t}^{t+\Delta}\sigma(y^{\varepsilon}(s),x_s^{\varepsilon},\frac{\varepsilon}{\sqrt{s}}))dW(s)$$

$$+\frac{1}{2}[\varphi_{yy}''(y,x,v)+\varphi_{yyv}'''(y,x,v)[\varepsilon^{-1}\frac{\alpha}{\sqrt{t}}G(y,x,\frac{\varepsilon}{\sqrt{t}}v)+\frac{v}{2t}]\Delta$$

$$+\varphi_{yyyy}^{IV}(y,x,v)a(y,x)\Delta]\int\limits_{t}^{t+\Delta}\sigma(y^{\varepsilon}(s),x_s^{\varepsilon},\frac{\varepsilon}{\sqrt{s}}))dW(s)]^2+o(\Delta).$$

This enables us to deduce the following relation for the first term in [6.27]:

$$\mathbb{E}_{y,x,v}[\varphi(y+\Delta y^{\varepsilon},x,v+\Delta v^{\varepsilon})-\varphi(y,x,v)](1-\varepsilon^{-2}q(x)\Delta+o(\Delta))$$

$$=\varphi_v'(y,x,v)[\varepsilon^{-1}\frac{\alpha}{\sqrt{t}}G(y,x,\frac{\varepsilon}{\sqrt{t}}v)+\frac{v}{2t}]\Delta$$

$$+\varphi_y'(y,x,v)a(y,x)\Delta+\frac{1}{2}\varphi_{yy}''(y,x,v)\sigma^2(y^{\varepsilon}(s),x,\frac{\varepsilon}{\sqrt{t}}v)\Delta+o(\Delta).$$

Similarly, we obtain the following relation for the second term in [6.27]:

$$\mathbb{E}_{y,x,v}[\varphi(y+\Delta y^{\varepsilon},x_{t+\Delta}^{\varepsilon},v+\Delta v^{\varepsilon})-\varphi(y,x,v)][\varepsilon^{-2}q(x)\Delta+o(\Delta)]$$

$$=\varepsilon^{-2}q(x)\mathbb{E}_{y,x,v}[\varphi(y,x_{t+\Delta}^{\varepsilon},v)-\varphi(y,x,v)]+o(\Delta).$$

Thus, according to the definition, for the generator [6.25], we obtain

$$\mathbf{L}_t^{\varepsilon}(x)\varphi(y,x,v):=\lim_{\Delta\to 0}\frac{1}{\Delta}\mathbb{E}_{y,x,v}[\varphi(y+\Delta y^{\varepsilon},x_{t+\Delta}^{\varepsilon},v+\Delta v^{\varepsilon})-\varphi(y,x,v)]$$

$$=\varepsilon^{-2}\mathbf{Q}\varphi(y,x,v)+\mathbf{V}_t^{\varepsilon}(x)\varphi(y,x,v),$$

where

$$\mathbf{V}_t^{\varepsilon}(x)\varphi(y,x,v)=\varphi_v'(y,x,v)[\varepsilon^{-1}\frac{\alpha}{\sqrt{t}}G(y,x,\frac{\varepsilon}{\sqrt{t}}v)+\frac{v}{2t}]$$

$$+\varphi_v'(y,x,v)a(x,y)+\frac{1}{2}\varphi_{yy}''(y,x,v)\sigma^2(y,x,\frac{\varepsilon}{\sqrt{t}}v).$$

LEMMA 6.5.– *On the test functions* $\varphi(y, x, v) \in C^{3,0,3}(R, X, R)$, *the generator* $\mathbf{L}_t^\varepsilon(x)$ *has the following asymptotic representation:*

$$\mathbf{L}_t^\varepsilon(x)\varphi(y, x, v) = \varepsilon^{-2}\mathbf{Q}\varphi(y, x, v) + \varepsilon^{-1}\frac{1}{\sqrt{t}}\mathbf{G}_0(y, x)\varphi(y, x, v)$$

$$+\frac{1}{t}\mathbf{V}(y, x)\varphi(y, x, v) + \mathbf{A}(y, x)\varphi(y, x, v) + o(\varepsilon), \quad [6.28]$$

where

$$\mathbf{G}_0(y, x)\varphi(y, x, v) = \alpha G(y, x, 0)\varphi'_v(y, x, v),$$

$$\mathbf{V}(y, x)\varphi(y, x, v) = v(\alpha G'_v(y, x, 0) + \frac{1}{2})\varphi'_v(y, x, v),$$

$$\mathbf{A}(y, x)\varphi(y, x, v) = +a(y, x)\varphi'_v(y, x, v) + \frac{1}{2}\sigma^2(y, x, 0)\varphi''_{yy}(y, x, v).$$

PROOF.– By using the decompositions

$$G(y, x, \frac{\varepsilon}{\sqrt{t}}v) = G(y, x, 0) + G'_v(y, x, 0)\frac{\varepsilon}{\sqrt{t}}v + o(\varepsilon),$$

$$\sigma^2(y, x, \frac{\varepsilon}{\sqrt{t}}v) = \sigma^2(y, x, 0) + o(\varepsilon),$$

we derive relation [6.27] from [6.26].

Consider a perturbed test function

$$\varphi_t^\varepsilon(y, x, v) = \varphi(y, v) + \varepsilon\frac{1}{\sqrt{t}}\varphi_1(y, x, v) + \varepsilon^2\frac{1}{t}\varphi(y, x, v).$$

LEMMA 6.6.– *The solution of the singular perturbation problem for the truncated generator*

$$\mathbf{L}_{t_0}^\varepsilon(x)\varphi(y, x, v) = \varepsilon^{-2}\mathbf{Q}\varphi(y, x, v) + \varepsilon^{-1}\frac{1}{\sqrt{t}}\mathbf{G}_0(y, x)\varphi(y, x, v)$$

$$+\frac{1}{t}\mathbf{V}(y, x)\varphi(y, x, v) + \mathbf{A}(y, x)\varphi(y, x, v) \quad [6.29]$$

on the test functions $\varphi_t^\varepsilon(y, x, v)$ *with* $\varphi(x, y) \in C^{3,3}(R \times R)$ *has the form*

$$\mathbf{L}_{t_0}^\varepsilon(x)\varphi_t^\varepsilon(y, x, v) = \frac{1}{t}\mathbf{L}\varphi(y, v) + \varepsilon\theta_t^\varepsilon(x)\varphi(y, v), \quad [6.30]$$

where the limit operator $\mathbf{L}$ *is given by the formula*

$$\mathbf{L}\varphi(y,v) = v(\alpha g(y) + \frac{1}{2})\varphi'_v(y,v) + \frac{1}{2}\alpha^2\sigma_v^2(y)\varphi''_{vv}(y,v)$$

$$+ t\hat{a}(y)\varphi'_y(y,v) + \frac{t}{2}\hat{\sigma}_y^2(y)\varphi''_{yy}(y,v), \qquad [6.31]$$

$$\hat{a}(y) = \int_X \pi(dx)a(y,x), \;\; \hat{\sigma}_y^2(y) = \int_X \pi(dx)\sigma^2(y,x,0).$$

PROOF.– According to the scheme of the solution of the singular perturbation problem (Korolyuk and Limnios 2005), we compute the value of generator [6.29] on the perturbed function [6.29] and on the perturbed function $\varphi_t^\varepsilon(y,x,v)$ as follows:

$$\mathbf{L}_{t_0}^\varepsilon(x)\varphi_t^\varepsilon(y,x,v) = \varepsilon^{-2}\mathbf{Q}\varphi(y,v) + \varepsilon^{-1}\frac{1}{\sqrt{t}}[\mathbf{Q}\varphi_1(y,x,v)$$

$$+\mathbf{G}_0(y,x)\varphi(y,v)] + \frac{1}{t}\mathbf{Q}\varphi_2(y,x,v) + \frac{1}{t}\mathbf{G}_0(y,x)\varphi_1(y,x,v)$$

$$+\frac{1}{t}(\mathbf{V}(y,x) + t\mathbf{A}(y,x))\varphi(y,v) + \varepsilon\theta_\varepsilon(x)\varphi,$$

where

$$\theta_t^\varepsilon(x)\varphi(y,v) = \mathbf{A}(y,x)\varphi_1(y,x,v) + \frac{1}{t^{3/2}}\mathbf{G}_0(y,x)\varphi_2(y,x,v)$$

$$+\varepsilon\frac{1}{t^2}\mathbf{V}(y,x)\varphi_2(y,x,v) + \varepsilon\mathbf{A}(y,x)\varphi_2(y,x,v).$$

Since $\mathbf{Q}\varphi(y,v) = 0$, the function $\varphi_1(y,x,v)$ satisfies the equation

$$\mathbf{Q}\varphi_1(y,x,v) + \mathbf{G}_0(y,x)\varphi_1(y,v) = 0.$$

By using the balance condition [6.22*], we obtain a solution of this equation in the form

$$\varphi_1(y,x,v) = \mathbf{R}_0\mathbf{G}_0(y,x)\varphi(y,v) = \alpha\mathbf{R}_0 G(y,x,0)\varphi'_v(y,v).$$

We now consider the equation for the function $\varphi_2(y,x,v)$, namely,

$$\mathbf{Q}\varphi_2(y,v,x) + \mathbf{G}_0(y,x)\varphi_1(y,v,x) + (\mathbf{V}(y,v) + t\mathbf{A}(y,v)\varphi = \mathbf{L}\varphi(y,v), \qquad [6.32]$$

where the limit operator $\mathbf{L}$ is determined from the condition of solving [6.31]:

$$\mathbf{L} = \Pi\mathbf{G}_0(y,x)\mathbf{R}_0\mathbf{G}_0(y,x) + \Pi\mathbf{V}(y,x) + t\Pi\mathbf{A}(y,x). \qquad [6.33]$$

Computing the right-hand side of [6.33], we obtain [6.31].

Equation [6.32] has the representation

$$\mathbf{Q}\varphi_2 + \mathbf{L}(y,x)\varphi(y,x) = \mathbf{L}\varphi(y,x),$$

where

$$\mathbf{L}(y,x) = \mathbf{G}_0(y,x) + \mathbf{R}_0\mathbf{G}_0(y,x) + (\mathbf{V}(y,x) + t\mathbf{A}(y,x)).$$

From the last representation and relation [6.31], we obtain

$$\varphi_2(y,x,v) = \mathbf{R}_0\tilde{L}(y,x)\varphi(y,v),$$

where $\tilde{L}(y,x) = \mathbf{L}(y,x) - \mathbf{L}$.

PROOF OF THEOREM 6.3.– From the smoothness of the components of system [6.18], [6.19], and the representations of the functions φ_1 and φ_2, we conclude that the remainder is bounded, i.e.

$$|\theta_t^\varepsilon(x)\varphi(y,v)| < M, \ M > 0. \tag{6.34}$$

The convergence of the processes $y^\varepsilon(t)$ and $v^\varepsilon(t)$ to the processes $\xi(t)$ and $\zeta(t)$ follows from [6.30] and [6.34] according to the Korolyuk model theorem (Korolyuk and Limnios 2005). Here, the generator of the process $\xi(t)$ has the form

$$\mathbf{L}_y\varphi(y,v) = t\hat{a}(y)\varphi_y'(y,v) + \frac{t}{2}\hat{\sigma}_y^2(y)\varphi_{yy}''(y,v).$$

The generator of the limit process $\zeta(t)$ has the form [6.31] and is the generator of the Ornstein–Uhlenbeck process.

6.2. Asymptotic properties of the impulse perturbation process with a control function under Levy approximation conditions

We construct limit generators of the impulse perturbation process and the dynamical system for the stochastic evolutionary equations with Markov switchings under Levy approximation conditions and at a single point of equilibrium of the quality criterion. In particular, we answer the question how the behavior of the boundary process depends on the prelimiting normalization of the stochastic evolutionary system in the ergodic Markov environment.

We consider some prelimiting evolution models with a small normalization parameter. We find the form of the limit generators for the impulse processes and the dynamical system in the schemes of the Levy approximation and the stochastic approximation. Furthermore, we provide the conditions that ensure weak convergence of a controlled evolution model with Markov switching and impulse perturbation (assuming uniqueness of the equilibrium point for the quality criterion for which the stochastic approximation procedure is given). It is important that the asymptotic behavior of the limit process is concluded with the help of the analysis of parameters of the prelimit system.

This work continues in the direction of research carried out in Korolyuk and Limnios (2005) as well as in Albeverio *et al.* (2008), Anisimov (2008) and Bainov and Simeonov (1992), in which the asymptotic properties of evolution models were obtained under the conditions of nonclassical approximation schemes.

The stochastic evolution system in the ergodic Markovian environment is given by a stochastic differential equation (Khimka and Chabanyuk 2013):

$$dy^{\varepsilon}(t) = C(y^{\varepsilon}(t), x(t/\varepsilon^2))dt + d\eta^{\varepsilon}(t, u^{\varepsilon}(t)), \qquad [6.35]$$

where $y^{\varepsilon}(t)$ is a random evolution and $x(t)$, $t \geq 0$ is a uniformly ergodic Markovian process that is defined on a standard phase space $(X, \mathbb{X})$ by the generator

$$Q\varphi(x) = q(x) \int_E P(x, dy)[\varphi(y) - \varphi(x)]$$

on the Banach space $B(X)$ of real-valued bounded functions with the supremum norm

$$\|\varphi(x)\| = \sup_{x \in X} |\varphi(x)|,$$

and $u^{\varepsilon}(t)$ is the control function, which we will define later.

The stochastic kernel $P(x, B)$, $x \in X$, $B \in \mathbb{X}$ defines the uniformly ergodic embedded Markov chain $x_n = x(\tau_n)$, $n \geq 0$ that has stationary distribution $\rho(B)$, $B \in \mathbb{X}$. The stationary distribution $\pi(B)$, $B \in \mathbb{X}$ is defined by the relation (Korolyuk and Korolyuk 1999)

$$\pi(dx)q(x) = q\rho(dx),$$

where

$$q = \int_X \pi(dx)q(x).$$

We define R_0 as the potential operator for the generator Q by the relation

$$R_0 = \Pi - (\Pi + Q)^{-1}.$$

Here, $\Pi\varphi(x) = \int_X \pi(dy)\varphi(y)1(x)$ is a projector onto the subspace $N_Q = \varphi : Q\varphi = 0$ of zeros to Q.

Under the Levy approximation scheme, the impulse perturbation process (IPP) $\eta^\varepsilon(t), t \geq 0$, is given by the relation

$$\eta^\varepsilon(t, u^\varepsilon(t)) = \int_0^t \eta^\varepsilon(ds, u^\varepsilon(s), x(t/\varepsilon^2)); \qquad [6.36]$$

where a set of processes with independent increments $\eta^\varepsilon(t, u, x)$, $t \geq 0$, $x \in X$, $u \in \mathbb{R}$ is determined by the generators

$$\Gamma^\varepsilon(x)\varphi(u, w) = \varepsilon^{-2} \int_{\mathbb{R}} (\varphi(u, w + v) - \varphi(u, w))\Gamma^\varepsilon(dv, x), \;\; x \in X \qquad [6.37]$$

and the following Levy approximation conditions are satisfied ([2,3]):

L1. Approximation of averages

$$\int_R v\Gamma^\varepsilon(dv, x) = \varepsilon a_1(x) + \varepsilon^2(a_2(x) + \theta_a(x)), \;\; \theta_a(x) \to 0, \;\; \varepsilon \to 0,$$

and

$$\int_R v^2\Gamma^\varepsilon(dv, x) = \varepsilon(b(x) + \theta_b(x)), \;\; \theta_b(x) \to 0, \;\; \varepsilon \to 0;$$

L2. Condition on the distribution function

$$\int_R g(v)\Gamma^\varepsilon(dv, x) = \varepsilon^2(\Gamma_g(x) + \theta_g(x)), \;\; \theta_g(x) \to 0, \;\; \varepsilon \to 0,$$

for each $g(v) \in C^2(\mathbb{R})$ (the space of bounded functions with values in $\mathbb{R}$ and $g(v)/|v|^2 \to 0$, $|v| \to 0$). The measure $\Gamma_g(x)$ is bounded for each $g(v) \in C^2(\mathbb{R})$ and can be determined by the relation

$$\Gamma_g(x) = \int_R g(v)\Gamma_0(dv, x), \;\; g(v) \in C^2(\mathbb{R});$$

L3. Uniform quadratic integrability

$$\sup \lim_{c \to \infty} \int_{|v|>c} v^2\Gamma_0(dv, x) = 0.$$

Let the criterion of the process quality be determined by the function $G(y, x, u)$ that has a unique point of equilibrium u_x^* in the interval $[\tau_i, \tau_{i+1}]$. It follows from the condition

$$G(y(t), x(t), u(t)) = 0. \qquad [6.38]$$

We note that the solution of the stochastic equation [6.35] on $[\tau_i, \tau_{i+1}]$ forms the Markov process with nonrandom control.

For problems [6.35] and [6.38], we consider the stochastic approximation procedure

$$du^\varepsilon(t) = \alpha(t)G(y^\varepsilon(t), x(t/\varepsilon^2), u^\varepsilon(t))dt. \qquad [6.39]$$

where we assume that the function $\alpha(t) > 0$ satisfies the following conditions:

P1: $\int\limits_{t_0}^{\infty} \alpha(t)dt = \infty$, $t_0 > 0$;

P2: $\int\limits_{t_0}^{\infty} \alpha^2(t)dt < \infty$, $t_0 > 0$.

Now we can formulate the main result of our paper.

THEOREM 6.4.– *Let the balance condition*

$$\hat{a}_1 := \int\limits_X \pi(dx)a_1(x) = 0 \qquad [6.40]$$

and L1–L3 hold.

Then, the weak convergence

$$(y^\varepsilon(t), u^\varepsilon(t), \eta^\varepsilon(t)) \Rightarrow (\hat{y}(t), \hat{u}(t), \hat{\eta}(t)), \ \varepsilon \to 0$$

holds true.

The limit process $(\hat{y}(t), \hat{u}(t), \hat{\eta}(t))$ *is determined by the generator*

$$\mathbf{M}\varphi(y, u, w) = \mathbf{L}\varphi(y, u, w) + \mathbf{B}\varphi(y, u, w), \qquad [6.41]$$

where

$$\mathbf{L}\varphi(y, u, w) = C(y)\varphi(y, u, w) + \Gamma\varphi(y, u, w),$$

$$C(y)\varphi(y) = \hat{C}(y)\varphi'(y), \ \hat{C}(y) = \int\limits_X \pi(dx)C(y, x),$$

$$\Gamma\varphi(u, w) = \hat{a}_2\varphi'_w(u, w) + \frac{1}{2}\sigma^2\varphi''_{ww}(u, w) + \int\limits_R [\varphi(u, w+v) - \varphi(u, w)]\hat{\Gamma}_0(dv),$$

$$\hat{a}_2 = \int\limits_X \pi(dx)(a_2(x) - a_0(x)), \ \sigma^2 = \int\limits_X \pi(dx)(b(x) - b_0(x))$$

$$+2\int_X \pi(dx)a_1(x)R_0a_1(x),$$

$$a_0(x) = \int_R v\Gamma_0(dv,x),\ b_0(x) = \int_R v^2\Gamma_0(dv,x),\ \hat{\Gamma}_0(v) = \int_X \Gamma_0(dv,x),$$

$$\mathbf{B}\varphi(y,u,w) = \alpha(t)G(y,u)\varphi'_u(y,u,w),$$

$$G(y,u) = \int_X G(y,x,u)\pi(dx).$$

PROOF OF THEOREM 6.4.–

First, we investigate some properties of generators with independent increments.

LEMMA 6.7.– *Under conditions L1–L3, generators with independent increments $\eta^\varepsilon(t,u,x),\ t \geq 0,\ x \in X,\ u \in \mathbb{R}$ can be represented as follows*

$$\Gamma^\varepsilon(x)\varphi(u,w) = \varepsilon^{-1}\Gamma_1(x)\varphi(u,w) + \Gamma_2(x)\varphi(u,w) \qquad [6.42]$$

on the test functions $\varphi(u,w) \in \mathbb{R}\times C_3(\mathbb{R})$.

Here

$$\Gamma_1(x)\varphi(u,w) = a_1(x)\varphi'_w(u,w),$$

$$\Gamma_2(x)\varphi(u,w) = (a_2(x) - a_0(x))\varphi'_w(u,w) + \frac{1}{2}(b(x) - b_0(x))\varphi''_{ww}(u,w)$$
$$+\int_R [\varphi(u,w+v) - \varphi(u,w)]\Gamma_0(dv,x).$$

PROOF.– We use the Taylor series of the function $\varphi(u,w)$ and the transform generator [6.37]:

$$\Gamma^\varepsilon(x)\varphi(u,w) = \varepsilon^{-2}\int_R (\varphi(u,w+v) - \varphi(u,w))\Gamma^\varepsilon(dv,x)$$

$$= \varepsilon^{-2}\int_R (\varphi(u,w+v) - \varphi(u,w) - v\varphi'_w(u,w) - \frac{1}{2}v^2\varphi''_{ww}(u,w))\Gamma^\varepsilon(dv,x)$$

$$+\varepsilon^{-2}\int_R v\varphi'_w(u,w)\Gamma^\varepsilon(dv,x) + \frac{\varepsilon^{-2}}{2}\int_R v^2\varphi''_{ww}(u,w))\Gamma^\varepsilon(dv,x)$$

$$= \int_R (\varphi(u,w+v) - \varphi(u,w) - v\varphi'_w(u,w) - \frac{1}{2}v^2\varphi''_{ww}(u,w))\Gamma^0(dv,x)$$

$$+\varepsilon^{-1}a_1(x)\varphi'_w(u,w)+a_2(x)\varphi'_w(u,w)+\frac{1}{2}b(x)\varphi''_{ww}(u,w)+\gamma^\varepsilon(x)\varphi(u,w)$$

$$=\varepsilon^{-1}a_1(x)\varphi'_w(u,w)+(a_2(x)-a_0(x))\varphi'_w(u,w)+\frac{1}{2}(b(x)-b_0(x))\varphi''_{ww}(u,w)$$

$$+\int_R(\varphi(u,w+v)-\varphi(u,w))\Gamma^0(dv,x)+\gamma^\varepsilon(x)\varphi(u,w),$$

where before the last equality follows from Levy conditions L1–L3.

We note that the function $\varphi(\cdot,w+v)-\varphi(\cdot,w)-v\varphi'_w(\cdot,w)-\frac{1}{2}v^2\varphi''_{ww}(\cdot,w+v)\in\mathbb{C}_3(\mathbb{R})$ since it is bounded because $\varphi(\cdot,w)$ is bounded with its derivatives at the variable w and

$$\lim_{|v|\to 0}\frac{[\varphi(u,w+v)-\varphi(u,w)-v\varphi'_w(u,w)-\frac{1}{2}v^2\varphi''_{ww}(u,w+v)]}{|v|^2}=0.$$

Remembering that $\gamma^\varepsilon(u,w)\varphi(u,w)=O(\varepsilon^2)$, we obtain [6.39]. Lemma 6.7 is proved.

LEMMA 6.8.– *The generator of a three-component Markov process* $(u^\varepsilon,\eta^\varepsilon,x)$, $t\geq 0$ *has the form*

$$\hat{\Gamma}^\varepsilon(x)\varphi(u,w,x)=\varepsilon^{-2}Q\varphi(u,w,x)+\varepsilon^{-1}\Gamma_1(x)\varphi(u,w,x)$$
$$+\Gamma_1(x)\varphi(u,w,x)+\gamma^\varepsilon(x)\varphi(u,w,x), \qquad [6.43]$$

where the operators $\Gamma_1(x)$, $\Gamma_2(x)$ *are determined in Lemma 6.7, and the remainder term* $\|\gamma^\varepsilon(x)\varphi(u,w,x)\|\to 0$ *as* $\varepsilon\to 0$, $\varphi(\cdot,w,\cdot)\in\mathbb{C}(\mathbb{R})$.

PROOF.– Lemma 6.8 becomes clear by using the definitions of the generator of the Markov process and the forms of the generators of η^ε, u^ε and x.

We introduce a truncated generator:

$$\Gamma_0^\varepsilon(x)\varphi(u,w,x)=\varepsilon^{-2}Q\varphi(u,w,x)+\varepsilon^{-1}\Gamma_1(x)\varphi(u,w,x)+\Gamma_1(x)\varphi(u,w,x). \qquad [6.44]$$

LEMMA 6.9.– *Under the balance condition [6.38], the solution of the singular perturbation problem for the truncated operator is determined by the relation*

$$\Gamma_0^\varepsilon(x)\varphi^\varepsilon(u,w,x)=\Gamma\varphi(u,w)+\varepsilon\theta_\eta^\varepsilon(x)\varphi(u,w) \qquad [6.45]$$

on the test functions

$$\varphi^\varepsilon(u,w,x)=\varphi(u,w)+\varepsilon\varphi_1(u,w,x)+\varepsilon^2\varphi_2(u,w,x),$$

and the remainder term $\theta^{\varepsilon}_{\eta}(x)\varphi(u,w,x)$ *is uniformly bounded at* x.

The limit operator is determined by the formula

$$\Gamma = \Pi\Gamma_1(x)R_0\Gamma_1(x)\Pi + \Pi\Gamma_2(x)\Pi. \qquad [6.46]$$

PROOF.– To perform equality [6.45], it is necessary that the coefficients of the same degrees of ε from left and right become equal. Let us calculate:

$$\begin{aligned}\Gamma^{\varepsilon}_0(x)\varphi^{\varepsilon}(u,w,x) &= \varepsilon^{-2}\mathbf{Q}\varphi(u,w) + \varepsilon^{-1}[\mathbf{Q}\varphi_1(u,w,x) + \Gamma_1(x)\varphi(u,w)] \\ &+ [\mathbf{Q}\varphi_2(u,w,x) + \Gamma_1(x)\varphi_1(u,w,x) + \Gamma_2(x)\varphi(u,w)] \\ &+ \varepsilon[\Gamma_1(x)\varphi_2(u,w,x) + \Gamma_2(x)\varphi_1(u,w,x)] \\ &+ \varepsilon^2\Gamma_2(x)\varphi_2(u,w,x).\end{aligned}$$

The first summation follows the relation

$$\mathbf{Q}\varphi(u,w) = 0 \Leftrightarrow \varphi(u,w) \in N_Q.$$

and we can see that $\varphi(u,w)$ does not depend on x.

The balance condition is a solvability condition of the equation

$$\mathbf{Q}\varphi_1(u,w,x) + \Gamma_1(x)\varphi(u,w) = 0.$$

Then

$$\varphi_1(u,w,x) = R_0\Gamma_1(x)\varphi(u,w). \qquad [6.47]$$

Let us consider the equation

$$\mathbf{Q}\varphi_2(u,w,x) + \Gamma_1(x)\varphi_1(u,w,x) + \Gamma_2(x)\varphi(u,w) = \Gamma\varphi(u,w).$$

It can be rewritten in the form

$$\mathbf{Q}\varphi_2(u,w,x) + \Gamma_1(x)R_0\Gamma_1(x)\varphi(u,w) + \Gamma_2(x)\varphi(u,w) = \Gamma\varphi(u,w).$$

From the solvability condition in the last equation, we can find the limit operator in the form [6.46]. Then

$$\varphi_2(u,w,x) = R_0[\Gamma_1(x)R_0\Gamma_1(x) + \Gamma_2(x) - \Gamma]\varphi(u,w). \qquad [6.48]$$

Using [6.47] and [6.48], another term of Taylor series can be written as follows:

$$\begin{aligned}&\varepsilon[\Gamma_1(x)\varphi_2(u,w,x) + \Gamma_2(x)\varphi_1(u,w,x)] + \varepsilon^2\Gamma_2(x)\varphi_2(u,w,x) \\ &= \varepsilon[\Gamma_1(x)R_0[\Gamma_1(x)R_0\Gamma_1(x) + \Gamma_2(x) - \Gamma] + \Gamma_2(x)R_0\Gamma_1(x)] \\ &\quad + \varepsilon\Gamma_2(x)R_0[\Gamma_1(x)R_0\Gamma_1(x) + \Gamma_2(x) - \Gamma]\varphi(u,w) = \varepsilon\theta^{\varepsilon}_{\eta}(x)\varphi(u,w).\end{aligned}$$

The limitation of $\theta^{\varepsilon}_{\eta}(x)\varphi(u,w)$ follows from the representation of the operators Γ_1, Γ_2 and R_0.

LEMMA 6.10.– *The generator of a four-component Markov process* $(y^\varepsilon, u^\varepsilon, \eta^\varepsilon, x)$, $t \geq 0$ *has the form*

$$\mathbf{L}^\varepsilon(x)\varphi(y,u,w,x) = \varepsilon^{-2}Q\varphi(y,u,w,x) + \Gamma^\varepsilon(x)\varphi(y,u,w,x) + \mathbf{C}(x)\varphi(y,u,w,x) + \mathbf{B}(x)\varphi(y,u,w,x) + \theta^\varepsilon_w(x)\varphi(y,u,w,x), \quad [6.49]$$

where $\Gamma^\varepsilon(x)$ *is a generator of set impulse perturbation processes [6.37],*

$$\mathbf{C}(x)\varphi(y,x) = C(y,x)\varphi'_y(y,x), \ \mathbf{B}(x)\varphi(u) = \alpha(t)G(y,x,u)\varphi'(u). \quad [6.50]$$

The remainder term $\|\theta^\varepsilon_w(x)\varphi(u,w,x)\| \to 0$ *as* $\varepsilon \to 0$.

Proof of Lemma 6.10 can be found in Chabanyuk *et al.* (2019).

LEMMA 6.11.– *The generator* $\mathbf{L}^\varepsilon(x)$ *in the case of the impulse perturbation process has the asymptotic representation*

$$\mathbf{L}^\varepsilon(x)\varphi(u,w,x) = \varepsilon^{-2}\mathbf{Q}\varphi(u,w,x) + \varepsilon^{-1}\Gamma_1(x)\varphi(u,w,x) + \Gamma_2(x)\varphi(u,w,x) + \mathbf{C}(x)\varphi(u,w,x) + \mathbf{B}(x)\varphi(u,w,x) + \hat{\theta}^\varepsilon_w\varphi(u,w,x) \quad [6.51]$$

where

$$\hat{\theta}^\varepsilon_w(x) = \gamma^\varepsilon + \theta^\varepsilon_w(x),$$

and $\Gamma_1(x)$ *and* $\Gamma_2(x)$ *are determined in Lemma 6.7.*

Remainder term $\|\hat{\theta}^\varepsilon_w(x)\varphi(u,w,x)\| \to 0$ *at* $\varepsilon \to 0$.

PROOF.– Proof of Lemma 6.11 is carried out using the representation of the operator [6.39] and the results of Lemma 6.10.

A truncated generator has the form

$$\mathbf{L}^\varepsilon_0(x)\varphi(u,w,x) = \varepsilon^{-2}Q\varphi(u,w,x) + \varepsilon^{-1}\Gamma_1(x)\varphi(u,w,x) + \Gamma_2(x)\varphi(u,w,x) + \mathbf{C}(x)\varphi(u,w,x) + \mathbf{B}(x)\varphi(u,w,x). \quad [6.52]$$

LEMMA 6.12.– *Under the balance condition [6.38], the solution of the singular perturbation problem for the truncated operator [6.52] is determined by the relation*

$$\mathbf{L}^\varepsilon_0(x)\varphi^\varepsilon(y,u,w,x) = \mathbf{L}\varphi(y,u,w) + \varepsilon^2\theta^\varepsilon_w(x)\varphi(y,u,w) \quad [6.53]$$

on the test functions

$$\varphi^\varepsilon(y,u,w,x) = \varphi(y,u,w) + \varepsilon\varphi_1(y,u,w,x) + \varepsilon^2\varphi_2(y,u,w,x),$$

and remainder term $\theta_w^\varepsilon(x)$ *is uniformly bounded at* x.

The limit operator is determined by the formula

$$\mathbf{L} = \Pi[\mathbf{C}(x) + \Gamma_1(x)R_0\Gamma_1(x) + \Gamma_2(x) + \mathbf{B}(x)]\Pi. \qquad [6.54]$$

PROOF.– To perform equality [6.51], it is necessary that the coefficients of the same degrees of ε from left and right become equal. We calculate:

$$\begin{aligned}
\mathbf{L}_0^\varepsilon(x)\varphi^\varepsilon(y,u,w,x) = {} & \varepsilon^{-2}\mathbf{Q}\varphi(y,u,w) \\
& + \varepsilon^{-1}[\mathbf{Q}\varphi_1(y,u,w,x) + \Gamma_1(x)\varphi(y,u,w)] \\
& + [\mathbf{Q}\varphi_2(y,u,w,x) + \Gamma_1(x)\varphi_1(y,u,w,x) \\
& + \Gamma_2(x)\varphi(y,u,w) + \mathbf{C}(x)\varphi(y,u,w) + \mathbf{B}\varphi(y,u,w)] \\
& + \varepsilon[\Gamma_1(x)\varphi_2(y,u,w,x) + \Gamma_2(x)\varphi_1(y,u,w,x) \\
& + \mathbf{C}(x)\varphi_1(y,u,w,x)] + \varepsilon^2[\Gamma_2(x)\varphi_2(y,u,w,x) \\
& + \mathbf{C}(x)\varphi_2(y,u,w,x) + \mathbf{B}\varphi_2(y,u,w,x)].
\end{aligned}$$

Since

$$\mathbf{Q}\varphi(y,u,w) = 0 \Leftrightarrow \varphi(y,u,w) \in N_Q,$$

then $\varphi(y,u,w)$ does not depend on x.

The balance condition is a solvability condition of the equation

$$\mathbf{Q}\varphi_1(y,u,w,x) + \Gamma_1(x)\varphi(y,u,w) = 0.$$

Then

$$\varphi_1(y,u,w,x) = R_0\Gamma_1(x)\varphi(y,u,w).$$

We rewrite the equation

$$\begin{aligned}
& \mathbf{Q}\varphi_2(y,u,w,x) + \Gamma_1(x)\varphi_1(y,u,w,x) + \Gamma_2(x)\varphi(y,u,w) \\
& + \mathbf{C}(x)\varphi(y,u,w) + \mathbf{B}(x)\varphi(y,u,w) = \mathbf{L}\varphi(y,u,w)
\end{aligned}$$

in the form

$$\mathbf{Q}\varphi_2(y,u,w,x) = [\mathbf{L} - \Gamma_1(x)R_0\Gamma_1(x) - \Gamma_2(x) - \mathbf{C}(x) - \mathbf{B}(x)]\varphi(y,u,w).$$

From the solvability condition in the last equation, we can find the limit operator in the form [6.54].

The proof of the theorem is completed using Theorem 6.3 in Korolyuk and Limnios (2005).

NOTE 6.1.– *Weak convergence of the process* $(y^\varepsilon(t), u^\varepsilon(t), \eta^\varepsilon(t)) \Rightarrow (\hat{y}(t), \hat{u}(t), \hat{\eta}(t))$, $\varepsilon \to 0$ *follows from the convergence of the corresponding generators under the condition of the compactness of a prelimited set of processes* $(y^\varepsilon(t), u^\varepsilon(t))$. *The corresponding theorems on the compactness of processes with independent increments in the Levy approximation scheme have been proved, in particular, in Korolyuk and Limnios (2005).*

THEOREM 6.5.– *We assume that conditions L1–L2, P1, and P2 hold true and there exists a Lyapunov function* $V(y, u)$ *of the averaging system*

$$\frac{du}{dt} = G(y, u)$$

satisfying the conditions of exponential stability

$$Y1: \; G(y, u)V'(y, u) \leq -cV(y, u),$$

the additional conditions for the Lyapunov function

$$Y2: \; |C(y, x)R_0[\tilde{G}(y, x, u)V_u'(y, u)]_u'| \leq c_1(1 + V(y, u))$$

and the following conditions:

$$|G(y, x, u)R_0[\tilde{C}(y, x)V_u'(y, u)]_u'| \leq c_2(1 + V(y, u)),$$

$$|\tilde{C}(y, x)R_0[\tilde{C}(y, x)V_u''(y, u)]_u'| \leq c_3(1 + V(y, u)),$$
$$|G(y, x, u)R_0[\tilde{G}(y, x, u)V_u'(y, u)]_u'| \leq c_4(1 + V(y, u)).$$

Then, $\forall \varepsilon > 0$, $\varepsilon \leq \varepsilon_0$, $\varepsilon_0 > 0$, *and the relation*

$$P\{\lim_{t\to\infty} u(t) = u^*\} = 1, \; \forall y$$

holds true.

PROOF.– For the perturbed Lyapunov function

$$V^\varepsilon(y, u, w, x) = V(y, u) + \varepsilon V_1(y, u, w, x) + \varepsilon^2 V_2(y, u, w, x),$$

using [6.54], we obtain the limiting presentation for the control

$$L_u^\varepsilon V^\varepsilon(y, u, w, x) = \mathbf{B}V(y, u) + \varepsilon\theta_u(x).$$

From conditions $Y1$ and $Y2$, we have the estimate

$$L'_u V^\varepsilon(y, u, w, x) \leq -c\alpha(t)V(y, u) + c^* \alpha^2(t)(1 + V(y, u)),$$

and then we use the result of the Nevelson–Khasminsky theorem (Nevelson and Khasminsky 1972). The proof of Theorem 6.5 is complete.

NOTE 6.2.– *The stochastic approximation procedure for control u gives the optimal value u_y for any IPP y.*

References

Albeverio, S., Koshmanenko, V.D., and Samoilenko, I.V. (2008). The conflict interaction between two complex systems: Cyclic migration. *Journal of Interdisciplinary Mathematics*, 11(2), 163–185.

Albeverio, S., Koroliuk, V.S., and Samoilenko, I.V. (2009). Asymptotic expansion of semi-Markov random evolutions. *Stochastics: An International Journal of Probability and Stochastic Processes*, 81(5), 477–502.

Anisimov, V.V. (2008). *Switching Processes in Queuing Models*. ISTE Ltd, London, and Wiley, New York.

Bainov, D. and Simeonov, P. (1992). *Integral Inequalities and Applications*. Kluwer Academic Publishers, Dordrecht.

Bellman, R. (1957). *Dynamic Programming (Dover Books on Computer Science)*. Princeton University Press, Princeton.

Bertoin, J. (1996). *Lévy Processes*. Cambridge University Press, Cambridge.

Billingsley, P. (1968). *Convergence of Probability Measures*. Wiley, New York.

Bryc, W. (1990). Large deviations by the asymptotic value method. In *Diffusion Processes and Related Problems in Analysis*, Vol. I, Pinsky, M.A. (ed.), 447–472, Birkhäuser, Boston.

Chabanyuk, Y., Koroliuk, V.S., and Limnios, N. (2007). Fluctuation of stochastic systems with average equilibrium point. *Comptes Rendus de l'Académie des Sciences*, Paris, Series I, 345, 405–410.

Chabanyuk, Y.M. (2004a). Asymptotic normality for a continuous procedure of stochastic approximation in a Markov medium. *Dopov. Nats. Akad. Nauk Ukr.*, A(5), 37–45.

Chabanyuk, Y.M. (2004b). Continuous procedure of stochastic approximation in a semi-Markov medium. *Ukrains'kyi Matematychnyi Zhurnal*, 56, 713–720.

Chabanyuk, Y.M. (2006). Asymptotic normality of a discrete procedure of stochastic approximation in a semi-Markov medium. *Ukrainian Mathematical Journal*, 58, 1616–1625.

Chabanyuk, Y.M. (2007). Stability of a dynamical system with semi-Markov switchings under conditions of diffusion approximation. *Ukrainian Mathematical Journal*, 59, 1441–1452.

Chabanyuk, Y.M., Nikitin, A.V., and Khimka, U.T. (2019). Asymptotic properties of the impulse perturbation process under Levy approximation conditions with the point of equilibrium of the quality criterion. *Matematychni Studii*, 52, 96–104.

Chernoff, H. (1952). Measure of asymptotic efficiency for tests of a hypothesis based on the sum of observations. *Annals of Mathematical Statististics*, 23(4), 493–655.

Cinlar, E., Jacod, J., Protter, P., and Sharpe, M.J. (1980). Semimartingale and Markov processes. *Zeitschrift für ahrscheinlichkeitstheorie und Verwandte Gebiete*, 54, 161–219.

Cole, J.D. (1951). On a quasi-linear parabolic equation occurring in aerodynamics. *The Quarterly of Applied Mathematics*, 9, 225–236.

Cramer, H. (1938). Sur un nouveau théorème-limite de la théorie des probabilités. *Actualités Scientifiques et Industrielles*, 736, 5–23.

Crandall, M.G. and Lions, P.L. (1983). Viscosity solutions of Hamilton-Jacobi equations. *Transactions of the American Mathematical Society*, 277, 1–42.

Davis, M.H.A. (1993). *Markov Models and Optimization*. Chapman & Hall/CRC Press, Boca Raton.

Dellacherie, C. and Meyer, P.-A. (1993). *Probabilities and Potential*. North-Holland Publishing, Amsterdam-New York.

Dembo, A. and Zeitouni, O. (1993). *Large Deviations Techniques and Applications*. Jones and Bartlett Publishers, Boston.

Deuschel, J.-D. and Stroock, D.W. (1989). *Large Deviations*. Academic Press, Boston.

Devooght, J. (1997). Dynamic reliability. *Advances in Nuclear Science and Technology*, 25, 215–278.

Doléans-Dade, C. and Meyer P.-A. (1970). Intégrales stochastiques par rapport aux martingales locales. Séminaire de Probabilités IV Université de Strasbourg. *Lecture Notes in Mathematics*, 124, 77–107.

Donsker, M.D. and Varadhan, S.R.S. (1975a). Asymptotic evaluation of certain Markov Process expectations for large time. I, *Communications on Pure and Applied Mathematics*, 27, 1–47.

Donsker, M.D. and Varadhan, S.R.S. (1975b). Asymptotic evaluation of certain Markov Process expectations for large time. II, *Communications on Pure and Applied Mathematics*, 28, 279–301.

Donsker, M.D. and Varadhan, S.R.S. (1975c). Asymptotic evaluation of certain Markov Process expectations for large time. III, *Communications on Pure and Applied Mathematics*, 29, 389–461.

Doob, J.L. (1990). *Stochastic Processes*. Reprint of the 1953 Original. John Wiley & Sons, New York.

Dupuis, P. and Ellis, R.S. (1997). *A Weak Convergence Approach to the Theory of Large Deviations*. Wiley, New York.

Ethier, S.N. and Kurtz, T.G. (1985). *Markov Processes: Characterization and Convergence*. Wiley, New York.

Feng, J. (1999). Martingale problems for large deviations of Markov Processes, *Stochastic Processes and Applications*, 81(2), 165–216.

Feng, J. and Kurtz, T.G. (2006). *Markov Processes: Characterization and Convergence*. American Mathematical Society, Providence.

Fleming, W.H and Soner, H.M. (2006). *Controlled Markov Processes and Viscosity Solutions*. Springer, New York.

Fleming, W.H and Souganidis, P.E (1986). A PDE approach to some large deviations problems. *Lectures in Applied Mathematics*, 23, 441–447. American Mathematical Society, Providence.

Freidlin, M.I. (1978). The averaging principle and theorems on large deviations. *Russian Mathematical Surveys*, 33, 117–176.

Fukushima, M. and Stroock, D.W (1986). Reversibility of solutions to martingale problems. *Adv. Math. (Suppl. Stud.)*, 9, 107–123.

Gorostiza, L. (1972). A central limit theorem for a class of D-dimensional random motions with constant speed. *Bulletin of the American Mathematical Society*, 78, 575–577.

Gorostiza, L. (1973a). The central limit theorem for random motions of D-dimensional Euclidean space. *Annals of Probability*, 1, 603–612.

Gorostiza, L. (1973b). An invariance principle for a class of D-dimensional polygonal random functions. *Transactions of the American Mathematical Society*, 177, 413–445.

Griego R. and Hersh, R. (1969). Random evolutions, Markov chains, and systems of partial differential equations. *Proceedings of the National Academy of Sciences*, 62, 305–308.

Hersh, R. (1974). Random evolutions: Survey of results and problems. *Rocky Mountain Journal of Mathematics*, 4, 443–477.

Hersh, R. (2003). The birth of random evolutions. *Mathematical Intelligencer*, 25, 53–60.

Hersh, R. and Papanicolaou, G. (1972). Non-commuting random evolutions, and an operator-valued Feynman–Kac formula. *Communications on Pure and Applied Mathematics*, 30, 337–367.

Hersh, R. and Pinsky, M. (1972). Random evolutions are asymptotically Gaussian. *Communications on Pure and Applied Mathematics*, 25, 33–34.

Hillen, T. (2001). Transport equations and chemosensitive movement. Habilitation Thesis, University of Tubingen, Tubingen.

Hopf, E. (1950). The partial differential equation $u_t + uu_x = \mu u_{xx}$. *Communications on Pure and Applied Mathematics*, 3, 201–230.

Jacod, J. and Shiryaev, A.N. (2003). *Limit Theorems for Stochastic Processes.* Springer-Verlag, Berlin.

Khimka, U.T. and Chabanyuk, Y.M. (2013). A difference stochastic optimization procedure with impulse perturbation. *Cybernetics and Systems Analysis*, 49, 768–773.

Kolesnik, A.D. (2001). Weak convergence of a planar random evolution to the Wiener process. *Journal of Theoretical Probability*, 14(2), 485–494.

Kolesnik, A.D. (2003). Weak convergence of the distributions of Markovian random evolutions in two and three dimensions. *Buletinul Academiei de Stiinte a Republicii Moldova. Matematica*, 3, 41–52.

Koroliuk, V.S. and Samoilenko, I.V (2014). Large deviations for random evolutions in the scheme of asymptotically small diffusion. *Modern Stochastics and Applications, Optimization and its Applications*, 90, 201–217.

Koroliuk, V.S., Limnios, N., and Samoilenko, I.V. (2010a). Levy approximation of impulsive recurrent process with Markov switching. *Theory of Probability and Mathematical Statistics*, 80, 15–23.

Koroliuk, V.S., Limnios, N., and Samoilenko, I.V. (2010b). Levy approximation of impulsive recurrent process with semi-Markov switching. *Theory of Stochastic Processes*, 16(32), 77–85.

Korolyuk, V.S. and Chabanyuk, Y.M. (2002), Stability of a dynamical system with semi-Markov switchings under conditions of stability of the averaged system. *Ukrainian Mathematical Journal*, 54, 239–252.

Korolyuk, V.S. and Korolyuk, V.V. (1999). *Stochastic Models of Systems.* Kluwer Academic Publishers, Dordrecht.

Korolyuk, V. and Limnios, N. (2005). *Stochastic Systems in Merging Phase Space.* World Scientific, Singapore.

Korolyuk, V.S. and Turbin, A.F. (1993). *Mathematical Foundations of the State Lumping of Large Systems.* Kluwer Academic Publishers, Dordrecht.

Kurtz, T.G. (1973). A limit theorem for perturbed operator semigroups with applications to random evolutions. *Journal of Functional Analysis*, 12, 55–67.

Kurtz, T.G. (1987). Martingale problems for controlled processes. *Lecture Notes in Control and Information Sciences*, 91, 75–90.

Kushner, H.J. (1990). *Weak Convergence Methods and Singular Perturbed Stochastic Control and Filtering Problems*. Birkhäuser, Boston.

Limnios N. and Oprişan, G. (2001). *Semi-Markov Processes and Reliability*. Birkhäuser, Boston.

Mogulskii, A.A. (1993). Large deviation for processes with independent increments. *Annals of Probability*, 21, 202–215.

Nashed, M.Z. (1976). *Generalized Inverses and Applications*. Academic Press, New York.

Nevelson, M.S. and Khasminsky, R.Z. (1972). *Stochastic Approximation and Recurrence Evaluation*. Nauka, Moscow.

Nikitin, A.V. (2015). Asymptotic properties of a stochastic diffusion transfer process with an equilibrium point of a quality criterion. *Cybernetics and System Analysis*, 4(51), 650–656.

Nikitin, A.V. (2018). Asymptotic dissipativity of stochastic processes with impulsive perturbation in the Levy approximation scheme. *Journal of Automation and Information Sciences*, 4(50), 205–211.

Nikitin, A.V. and Khimka, U.T. (2017). Asymptotics of normalized control with Markov switchings. *Ukrainian Mathematical Journal*, 8(68), 1252–1262.

O'Brien, G.L. and Vervaat, W. (1995). Compactness in the theory of large deviations. *Stochastic Processes and Applications*, 57, 1–10.

Orsingher, E. (2002). Bessel functions of third order and the distribution of cyclic planar motions with three directions. *Stochastics and Stochastics Reports*, 74, 617–631.

Orsingher, E. and Sommella, A.M. (2004). A cyclic random motion in $\mathbf{R}^3$ with four directions and finite velocity. *Stochastics and Stochastics Reports*, 76, 113–133.

Othmer, H.G. and Hillen, T. (2002). The diffusion limit of transport equations II: Chemotaxis equations. *SIAM Journal on Applied Mathematics*, 4(62), 1222–1250.

Papanicolaou, G. (1971a). Motion of a particle in a random field. *Journal of Mathematical Physics*, 12, 1494–1496.

Papanicolaou, G. (1971b). Wave propagation in a one-dimensional random medium. *SIAM Journal on Applied Mathematics*, 21, 13–18.

Papanicolaou, G. (1973). Stochastic equations and their applications. *American Mathematical Monthly*, 80, 526–544.

Papanicolaou, G. (1975). Asymptotic analysis of transport processes. *Bulletin of the American Mathematical Society*, 81, 330–392.

Papanicolaou, G. (1976). Probabilistic problems and methods in singular perturbations. *Rocky Mountain Journal of Mathematics*, 6, 653–673.

Papanicolaou, G. and Hersh, R. (1972). Some limit theorems for stochastic equations and applications. *Indiana University Mathematics Journal*, 21, 815–840.

Papanicolaou, G. and Kellek, J. (1971). Stochastic differential equations with applications to random harmonic oscillations and wave propagation in random media. *SIAM Journal on Applied Mathematics*, 21, 287–205.

Papanicolaou, G., Stroock D., and Varadhan, R.S.S. (1977). Martingale approach to some limit theorems. In *Statistical Mechanics Dynamical Systems*, Ruelle D (ed.). Duke University, Durham.

Pinsky, M. (1968). Differential equations with a small parameter and the central limit theorem for functions defined on a finite Markov chain. *Zeitschrift fÃ¼r ahrscheinlichkeitstheorie und Verwandte Gebiete*, 2, 101–111.

Pinsky, M. (1991). *Lectures on Random Evolutions*. World Scientific, Singapore.

Protter, P. (1990). *Stochastic Integration and Differential Equations. A New Approach*. Springer-Verlag, Berlin.

Puhalskii, A. (1991). On functional principle of large deviations. In *New Trends in Probability and Statistics*. VSP, Utrecht, 198–219.

Roth, J.P. (1974). Opérateurs carré du champ et formule de lévy-khinchine sur les espaces localement compacts. *Comptes Rendus de l'Académie des Sciences Paris, Series A*, 278, 1103–1106.

Samoilenko, I.V. and Nikitin, A.V. (2018). Differential equations with small stochastic terms under the levy approximating conditions. *Ukrainian Mathematical Journal*, 69(9), 1445–1454.

Samoilenko, I.V. and Nikitin, A.V. (2019). Double merging of the phase space for stochastic differential equations with small additions in poisson approximation conditions. *Cybernetics and System Analysis*, 55(2), 265–273.

Samoilenko, I.V., Chabanyuk, Y.M., Nikitin, A.V., and Khimka, U.T. (2017). Differential equations with small stochastic additions under Poisson approximation conditions. *Cybernetics and System Analysis*, 53(3), 410–416.

Skorokhod, A.V. (1989). *Asymptotic Methods in the Theory of Stochastic Differential Equations*. American Mathematical Society, Providence.

Skorokhod, A.V., Hoppensteadt, F.C., and Salehi, H. (2002). *Random Perturbation Methods with Applications in Science and Engineering*. Springer-Verlag, New York.

Stroock, D.W. and Varadhan, R.S.S. (1979). *Multidimensional Diffusion Processes*. Springer-Verlag, Berlin.

Sviridenko, M.N. (1998). Martingale characterization of limit distributions in the space of functions without discontinuities of second kind. *Mathematical notes of the Academy of Sciences of the USSR*, 43(5), 398–402.

Index

Other titles from

iSTE

in

Mathematics and Statistics

2020

KOROLIOUK Dmitri
Dynamics of Statistical Experiments

MANOU-ABI Solym Mawaki, DABO-NIANG Sophie, SALONE Jean-Jacques
Mathematical Modeling of Random and Deterministic Phenomena

2019

BANNA Oksana, MISHURA Yuliya, RALCHENKO Kostiantyn, SHKLYAR Sergiy
Fractional Brownian Motion: Approximations and Projections

GANA Kamel, BROC Guillaume
Structural Equation Modeling with lavaan

KUKUSH Alexander
Gaussian Measures in Hilbert Space: Construction and Properties

LUZ Maksym, MOKLYACHUK Mikhail
Estimation of Stochastic Processes with Stationary Increments and Cointegrated Sequences

MICHELITSCH Thomas, PÉREZ RIASCOS Alejandro, COLLET Bernard, NOWAKOWSKI Andrzej, NICOLLEAU Franck
Fractional Dynamics on Networks and Lattices

VOTSI Irene, LIMNIOS Nikolaos, PAPADIMITRIOU Eleftheria, TSAKLIDIS George
Earthquake Statistical Analysis through Multi-state Modeling
(Statistical Methods for Earthquakes Set – Volume 2)

2018

AZAÏS Romain, BOUGUET Florian
Statistical Inference for Piecewise-deterministic Markov Processes

IBRAHIMI Mohammed
Mergers & Acquisitions: Theory, Strategy, Finance

PARROCHIA Daniel
Mathematics and Philosophy

2017

CARONI Chysseis
First Hitting Time Regression Models: Lifetime Data Analysis Based on Underlying Stochastic Processes
(Mathematical Models and Methods in Reliability Set – Volume 4)

CELANT Giorgio, BRONIATOWSKI Michel
Interpolation and Extrapolation Optimal Designs 2: Finite Dimensional General Models

CONSOLE Rodolfo, MURRU Maura, FALCONE Giuseppe
Earthquake Occurrence: Short- and Long-term Models and their Validation
(Statistical Methods for Earthquakes Set – Volume 1)

D'AMICO Guglielmo, DI BIASE Giuseppe, JANSSEN Jacques, MANCA Raimondo
Semi-Markov Migration Models for Credit Risk
(Stochastic Models for Insurance Set – Volume 1)

GONZÁLEZ VELASCO Miguel, del PUERTO GARCÍA Inés, YANEV George P.
Controlled Branching Processes
(Branching Processes, Branching Random Walks and Branching Particle Fields Set – Volume 2)

HARLAMOV Boris
Stochastic Analysis of Risk and Management
(Stochastic Models in Survival Analysis and Reliability Set – Volume 2)

KERSTING Götz, VATUTIN Vladimir
Discrete Time Branching Processes in Random Environment
(Branching Processes, Branching Random Walks and Branching Particle Fields Set – Volume 1)

MISHURA YULIYA, SHEVCHENKO Georgiy
Theory and Statistical Applications of Stochastic Processes

NIKULIN Mikhail, CHIMITOVA Ekaterina
Chi-squared Goodness-of-fit Tests for Censored Data
(Stochastic Models in Survival Analysis and Reliability Set – Volume 3)

SIMON Jacques
Banach, Fréchet, Hilbert and Neumann Spaces
(Analysis for PDEs Set – Volume 1)

2016

CELANT Giorgio, BRONIATOWSKI Michel
Interpolation and Extrapolation Optimal Designs 1: Polynomial Regression and Approximation Theory

CHIASSERINI Carla Fabiana, GRIBAUDO Marco, MANINI Daniele
Analytical Modeling of Wireless Communication Systems
(Stochastic Models in Computer Science and Telecommunication Networks Set – Volume 1)

GOUDON Thierry
Mathematics for Modeling and Scientific Computing

KAHLE Waltraud, MERCIER Sophie, PAROISSIN Christian
Degradation Processes in Reliability
(Mathematial Models and Methods in Reliability Set – Volume 3)

KERN Michel
Numerical Methods for Inverse Problems

RYKOV Vladimir
Reliability of Engineering Systems and Technological Risks
(Stochastic Models in Survival Analysis and Reliability Set – Volume 1)

2015

DE SAPORTA Benoîte, DUFOUR François, ZHANG Huilong
Numerical Methods for Simulation and Optimization of Piecewise Deterministic Markov Processes

DEVOLDER Pierre, JANSSEN Jacques, MANCA Raimondo
Basic Stochastic Processes

LE GAT Yves
Recurrent Event Modeling Based on the Yule Process
(Mathematical Models and Methods in Reliability Set – Volume 2)

2014

COOKE Roger M., NIEBOER Daan, MISIEWICZ Jolanta
Fat-tailed Distributions: Data, Diagnostics and Dependence
(Mathematical Models and Methods in Reliability Set – Volume 1)

MACKEVIČIUS Vigirdas
Integral and Measure: From Rather Simple to Rather Complex

PASCHOS Vangelis Th
Combinatorial Optimization – 3-volume series – 2nd edition
Concepts of Combinatorial Optimization / Concepts and Fundamentals – volume 1
Paradigms of Combinatorial Optimization – volume 2
Applications of Combinatorial Optimization volume 3

2013

COUALLIER Vincent, GERVILLE-RÉACHE Léo, HUBER Catherine, LIMNIOS Nikolaos, MESBAH Mounir
Statistical Models and Methods for Reliability and Survival Analysis

JANSSEN Jacques, MANCA Oronzio, MANCA Raimondo
Applied Diffusion Processes from Engineering to Finance

SERICOLA Bruno
Markov Chains: Theory, Algorithms and Applications

2012

BOSQ Denis
Mathematical Statistics and Stochastic Processes

CHRISTENSEN Karl Bang, KREINER Svend, MESBAH Mounir
Rasch Models in Health

DEVOLDER Pierre, JANSSEN Jacques, MANCA Raimondo
Stochastic Methods for Pension Funds

2011

MACKEVIČIUS Vigirdas
Introduction to Stochastic Analysis: Integrals and Differential Equations

MAHJOUB Ridha

Recent Progress in Combinatorial Optimization – ISCO2010

RAYNAUD Hervé, ARROW Kenneth
Managerial Logic

2010

BAGDONAVIČIUS Vilijandas, KRUOPIS Julius, NIKULIN Mikhail
Nonparametric Tests for Censored Data

BAGDONAVIČIUS Vilijandas, KRUOPIS Julius, NIKULIN Mikhail
Nonparametric Tests for Complete Data

IOSIFESCU Marius *et al.*
Introduction to Stochastic Models

VASSILIOU PCG
Discrete-time Asset Pricing Models in Applied Stochastic Finance

2008

ANISIMOV Vladimir
Switching Processes in Queuing Models

FICHE Georges, HÉBUTERNE Gérard
Mathematics for Engineers

HUBER Catherine, LIMNIOS Nikolaos *et al.*
Mathematical Methods in Survival Analysis, Reliability and Quality of Life

JANSSEN Jacques, MANCA Raimondo, VOLPE Ernesto
Mathematical Finance

2007

HARLAMOV Boris
Continuous Semi-Markov Processes

2006

CLERC Maurice
Particle Swarm Optimization